ENVIRONMENTAL IMPACT OF LARGE RESERVOIR PROJECTS ON HUMAN SETTLEMENT

Environmental Impact of Large Reservoir Projects on Human Settlement

(A Case Study of Upper Kolab Project in Orissa)

A.K. Dalua

ASHISH PUBLISHING HOUSE
8/81, Punjabi Bagh, New Delhi-110 026

Published by:
S.B. Nangia
for **Ashish Publishing House**
8/81, Punjabi Bagh,
New Delhi-110 026
Tel.: 500581, 5410924

Showroom : 5, Ansari Road, Delhi 110 002
Tel : 3274050

ISBN 81-7024-568-0

1993

Laser Typeset by:
Westvill Enterprises
Block B-5, Bungalow No. 47
Paschim Vihar,
New Delhi-110 063
Tel.: 558-7469

Printed at:
Efficient Offset Printers
Naraina, Delhi

Contents

1

Introduction

1

Introduction

(A) The Study

1. Any author of a work on resettlement, especially involuntary resettlement of displaced persons is under no delusion that now-a-days there is a great deal of general interest in the subject. Based on the causes, resettlement and rehabilitation activities can be broadly grouped into two categories:

(a) Voluntary and (b) Involuntary. Voluntary rehabilitation may be the outcome of displacement of human settlement either due to (i) natural calamities like earthquake, floods, cyclones and famines etc. or due to (ii) political, racial and religious disturbances etc. On the other hand, involuntary rehabilitation may become the result of displacement due to construction of various types of development projects.

2. Our Constitution, in its Seventh Schedule provides for the resettlement of persons uprooted; *i.e.*, the refugees consequent upon partition of the country. But there is no provision in the Constitution for the settlement of project displaced persons at whose cost the nation has marched ahead so far, with development activities. Natural calamities could not have been either foretold or controlled satisfactorily. But when the builders and architects of free India launched the Five-Year Plans, should have visualised the consequences of large scale displacement due to construction of various development projects and provided for orderly and ready to adopt guidelines for resettlement of these oustees.

3. The area of study of this particular disciplins in social science is relatively unexplored not only in India, but also

anywhere else in the advanced countries. As a matter of fact, the Western authorities have been experimenting various parameters of resettlement and rehabilitation measures in the World Bank Aided Projects in Africa and South-East Asian countries.

4. There may be many publications on resettlement and rehabilitation of oustees due to construction of various types of development projects. But so far as river valley projects are concerned; especially in Orissa, such type of study and publications are very negligible. Construction of the Rengali Multipurpose Project on the river Brahmani and the aftermath of resettlement activities of its oustees have attracted the attention of critics in different forums. But excepting some isolated works touching upon the socio-economic aspects, made by the teaching staff and students of Anthropology Department of Utkal University led by Dr. N.K. Behura, Professor and Dr. P.K. Nayak, Professor no systematic attempt seems to have been made so far to explain the real state of affairs. Similar isolated study has also been made by some of the students of aforesaid department, on World Bank Aided Upper Indravati Project which is now under construction. I.U. Bhaskara Reddy of the Tata Institute of Social Science, Bombay is understood to have done some work on Upper Kolab Project in 1983, by which time physical displacement of the denizens had not commenced. May be there are many other scholars who have done/are doing some work on this project. But the author is not aware of the same.

5. As is known to this author, the following are some of the good studies made by the Indian authors on the displacement and rehabilitation of river valley projects affected people in India:

(a) "Human Resettlement in Lower Narmada basin" by S.C. Verma, Chairman of the Naramada Valley Development Authority.

The study by Verma is more of an Approach paper to tackle the problem of displacement and resettlement than an overview of the logistic practical experiences in the field; because of the fact that real problems consequent to displacement is yet to commence.

(b) "Displacement and Rehabilitation" by Dr. B.L. Joshi of the Department of Sociology in Marathwada University, Aurangabad. Dr. Joshi did his doctoral thesis in 1979 on the oustees and their problems etc., arising out of construction of Jayakwadi Irrigation Project (Nathsagar) in Godabari basin in Maharashtra.

(c) "Loktak Project and some basic Issues for the Communities in Manipur Valley" by B.M. Mukerjee, Associate Professor, Department of Anthropology, Manipur University, Imphal. The observations made by Mukerjee are mostly based on anthropological field work conducted by him in 1987.

(d) "River Dam Construction and Resettlement of Affected villages" by Prof. Prakash R. Sawant of Department of Geography in Kisan Veer Mahavidyalaya, Wai (Dist: Satara) Maharashtra. Prof. Sawant has done the aforesaid study on rehabilitation of people affected by Dhom Dam Irrigation Project in Upper Krishna Valley, Maharashtra.

(e) "Planned Rehabilitation and Economic Change" by C.S. Gajarajan of Gokhale Institute of Politics and Economics, Pune. Gajarajan had made a case study on the oustees of Tungabhadra River Project.

(f) "Survey of the People Displaced Through the Koyna Dam" by Karve, Irawati and Jai Nimbakar of Deccan College, Pune.

6. Unlike Gujarat, Uttar Pradesh and Karanataka, there are no notable activists organisation in Orissa to champion the cause of the oustees. Orissa enjoys the pride of constructing the first multipurpose project of free India, in Hirakud in the fiftees. But even at this length of time, there are several unsettled land acquisition cases relating to this project.

7. To work on the present topic was felt by the author as an inner necessity while working in the State Government in erstwhile Irrigation and Power Department for about five years and dealing with intricate issues of planning, programming and monitoring the progress of execution of multipurose, major and medium irrigation projects in the State. During this period he came in intimate contact with all relevant activities including resettlement and rehabilitation of outsees of these projects. As a matter of fact, the urge to do some work in human settlement relating to project affected people occurred in the author when he visited Rengali Multipurpose Project area in Orissa in 1987 to review the measures taken by the project authorities to resettle the outsees of that Project. Subsequent to this, the author participated in a Seminar on "Water Resources Development - Environmental issues and Solutions," sponsored by the Government of India and United States Agency for International Development Group of Water and Power Consultancy Services (India) Ltd. at Bhubaneswar on 9 - 10th November 1987. He got an opportunity to visit the U.S.A. for a short training course on "Social and Technical Aspects of Irrigation Organisations" in Colorado State University. He also visited the Sardar Sarovar Project in Gujarat to obtain some first hand knowledge about the situation obtaining there. Thus, he experienced in his own person, what the trauma is there for a displaced person in losing his ancestral abode and in moving to an unknown place and take to avocations whether to his liking or not. He is of the opinion that every one who is in a position to do so should add his contribution to the knowledge of the evacuation processes and the life in resettlement colonies set up for them.

8. The importance given by the planners to the irrigation sector needs no emphasis. A close study of the planned development would reveal that a large part of the State's outlay has been in the field of irrigation and power. Again a half of it is for irrigation sector. While it was Rs. 315.30 crores only during the Plan period from 1951-52 to 1979-80, the investment during the two plan periods from 1980-81 to 1989-90 is Rs. 1105.00 crores. The outlay for Major and medium Ir-

rigation Sector alone during 1990-91 and 1991-92 was Rs. 160.54 crores and Rs. 188.20 crores respectively. The Eighth Plan outlay under this sector is of the order of Rs. 2614.33 cr. The accepted policy of the State Government is to maximise creation of irrigation potential especially to make the benefit reach the water shortage pockets in tribal and economically backward areas. Fortunately Orissa is richly endowed with water resources. Power generation in the State is largely dependent on hydro-electric projects. Power being an essential element in the growth of industries, hydro-electric projects also assume great importance.

9. Thus, the abundance of water resources while providing scope for development, also result some times in destruction of property and life due to the ravages of flood. Therefore, flood control also becomes an important part of development. It is in this context that irrigation projects including multipurpose projects have come to assume primary importance in the development plans in Orissa. But when one thinks about the magnitude of loss caused to the environment and displacement of human settlement due to submergence by the man-made reservoir, one gets shocked. The problem of rehabilitation of the displaced persons, more so in respect of irrigation projects has certain peculiar characteristics. It is not a problem merely relating to mechanical or inanimate resources, but rather a sensitive problem dealing with human beings, loss of forest wealth and wild life and above all, damage to the environment can never be properly measured as these are social costs, but nevertheless, they are very real. The problem becomes compounded by the fact that in many cases it touches the weakest sections of the population like the tribal population. It also involves special issues because, by and large the benefits of these projects are rarely available to the people who are adversely affected due to construction of these projects. Thus, the problem is to be dealt with more carefully and with a lot of sensitivity.

(B) Scope of the Study

1. It would be clear from the foregoing paragraphs that resettlement and rehabilitation of oustees is not only a hu-

man problem to be viewed with compassion, but also is a complex social system to be studied carefully. The author has tried to study this problem analytically in terms of the responsibility of Government organisations, voluntary organisations in particular and society in general.

2. The aim of this work is neither a warning to humanity, nor of any discovery; but rather to provide an explanation for what happened in the Upper Kolab Project in general and resettlement colonies set up for the oustees in particular, to the best of his ability. The author realises that he is under certain disadvantages for the fulfilment of so difficult a task; firstly because he is a government employee; and secondly because he is a generalist, without having any specialised training. But he is sure that with his long experience in the Administrative Service, he would be able to present the problem of human resettlement and suggest some measures to overcome them. The observations made by him are the outcome purely of his own assessment of the situation without any bias whatsoever. Critical observations wherever made are not with any intention to the disadvantage of anybody. They are only aimed at pointing outflaws, if any, so that they can be remedied to suit the greater interest of humanity.

3. Notwithstanding the ecological and socio-cultural variations not only in India but elsewhere in the world, the present study intends to focus the traumatic experiences and near insoluable plight of the oustees who are tribals and ruralites.

4. The study also intends to show that construction of major river valley dams cause colossal damage to the ecosystem and bring unmitigated miseries to the oustees if an integrated approach is not made in true sense of the terms. The issue of resettlement of the evacuees must be taken up long before the commencement of the construction of the project.

5. People's voluntary participation must be secured which is lacking not only in India but in other developing countries. This study highlights this problem.

6. This study presents a perspective of a person who has participated in the bureaucratic process that dealt with construction of dams in Orissa.

7. Another object of this study is to remove the idea of resistance from the mind of the oustees as well as the people in general to the construction of large and multipurpose projects. The popular view is that projects of such magnitude not only take long gestation period resulting thereby in cost overrun, but also cause irreparable human displacement. This view is conspicuous when an incipient agitation was staged against the construction of Rengali Multipurpose Project. A lot of publications are also there in the press about apprehended environmental losses due to Sardar Sarovar Project in Gujarat and Tehri Dam Project in Uttar Pradesh. But there are many brighter aspects of the Projects and it cannot be said that given adequate finance and support of right personnel, resettlement activities can be made most efficient to meet the requirement of the resettlers. The author has tried to place the facts, as objectively as he could to weigh the pros and cons of the subject in study. To the best of his knowledge, this is the first time that such a study on resettlement and rehabilitation issues of oustees of water resources projects in Orissa has been attempted. As such there is wide scope for further research.

(C) Methodology

1. Although the study discusses in general the rehabilitation policy relating to water resources projects of several other States, it is more a micro level study confined to a particular project *i.e.,* the Upper Kolab Multipurpose Project in Orissa. It cuts across many fields - environmental, socio-economic cultural, organisational, etc. The study is mostly the outcome of the personal experience of the author in course of routine administrative works; without much of reference to works done by emient scholars in this field.

2. In the first stage the project office formed the base for the collection of primary data. In this regard the Chief Engineer of the project and his staff, and the Special Land Acquisition and Resettlement Officer rendered ungrudging help.

3. In the second stage, data was collected from all the 424 oustee families resettled by November 1989, in the project sponsored colonies, through a structured questionnaire devised by the author himself.

4. Thirdly the author obtained first hand knowledge about the project, resettlement and resettlers by personal interview on random basis during his several visits to the project area. Besides, the materials contained in the various official records, committee reports, circulars and orders issued by Government of India as well as the State Government; recommendations of the Rehabilitation Advisory Committee of the Upper Kolab Project and other documents were collected and used at appropriate places.

(D) Structure

1. The work consists of five Chapters, besides the introductory Chapter, bearing Number 1.

2. The Second Chapter deals with the general setting of Orissa. It focuses attention mainly on the physical features of Orissa, its population and economy. Viewing that irrigation projects have a major role to play in influencing the economy of the people, there is a discussion on the need for irrigation projects.

3. The Third Chapter gives an idea about the characteristics of different categories of irrigation projects and some data about the volume of human displacement and the manner of their resettlement in some of the major and medium irrigation projects in Orissa.

4. Chapter Four presents the case in brief for a national policy on resettlement and rehabilitation, policy of the State Government *vis-a-vis* few other State Governments and World Bank guidelines on their involvement in project execution.

5. Chapter Five contains a detailed discussion on the Upper Kolab Project— its history, profile, cost, mode of land acquisition, procedure adopted in determining value of the land and property acquired, different phases of displacement of affected families and the manner of their resettlement.

6. The concluding Chapter *i.e.*, Chapter Six is in the nature of an enquiry into different problems of involuntary resettlement. It discusses the various lessons learnt from the displacement of the oustees of Upper Kolab Project and the arrangements made by the project authorities for their resettlement . It is divided into nine sections. Section 'A' gives an idea in brief about the lessons learnt from similar projects in Orissa and other States. Section 'B' gives an account of the many facets of environmental problems arising out of the project. Section 'C' discusses the different legal aspects connected with land acquisition, enumeration of displaced families and payment of compensation money as well as rehabilitation cash grant. Section 'D' contains a detailed discussion on the socio-economic issues relating to the resettlers. Section 'K', 'F', 'G' and 'H' deals with cultural, human, financial and organisational issues respectively. At the end of each section some suggestions have been offered to overcome the problems already confronted/apprehended to be confronted by the resettlers themselves and also by the administration. These suggestions are limited in application to the Upper Kolab Project only as the problems identified are related to them. Of course, they have no bearing on other projects; but no attempt has been made by the author to analyse them in that perspective, as the study is limited to the situation already prevailing in the resettlement colonies of Upper Kolab Project and that can be done to bring in improvements thereon. The concluding section 'I' contains important findings and some suggested guidelines based on this study.

There are repetitions at places which could not be avoided.

Sufficient and reliable data are essential prerequisites for proper planning in any sphere. Unfortunately detailed data with regard to submergence, displacement of human beings and their effective resettlement in Orissa are either very meagre, or there has been no proper documentation. There are, therefore, some gaps and certainly many new issues may need to be explored and tackled. What is missing in this study is what the author has not yet attempted or been able to do.

2

General Setting of Orissa

2

General Setting of Orissa

Introduction

Orissa, known as 'Kalinga' in anicent times, has a glorious past of being ruled by powerful kings who had extended its boundary from the Ganges in the North to the river Godavari in the South. Art and culture in its distinct form of Orissan style of temple architecture and "Odissi" dance and music flourished in this land under the patronage of these kings. The Oriya "sadhabas" (tradesmen) who had also pushed their trades and commerce activities into the far eastern countries, notable of them being Java, Sumatra and Borneo brought immense fortune. The golden days were over when in the 15th century, the kings devoted most of their time on religious and literary activities by neglecting development of muscle power and the "Pathans" and "Marhattas" took opportunity to invade Orissa. Ultimately in 1803, the British took over the administration. Until creation of the State of Bihar in 1912, Orissa along with Bihar remained as a part of greater Bengal Presidency. Subsequently upto 1936, it continued as a part of Bihar. In 1936, Orissa got the status of a separate State with six districts and twenty-six number of princely States. These princely States were merged in 1948 to form the modern Orissa State of today with thirteen number of districts.

Physiography

Orissa is a maritime State being located in the South-Eastern part of Inia between 17°-31° and 22°-27° North latitude and between 81°-27° and 87°-30° East longitude. It is bounded by Bihar in the North, West Bengal in the North-East, Madhya Pradesh in the West, Andhra Pradesh in the South-West and the Bay of Bengal in the East.

There are four well defined physical regions, namely the northern plateau (districts of Mayurbhanj, Keojhar and Sundargarh), Central Table land (districts of Bolangir, Dhenkanal, Sambalpur), Eastern ghats (districs of Koraput, Kalahandi and Phulbani) and the coastal plains(districts of Cuttack, Puri, Ganjam and Balasore). The coastline along the Bay of Bengal is about 480 Km long. Drawing No. 1 gives a picture of Orissa *vis-a-vis* location of Upper Kolab Project and some other major and medium irrigation projects.

The soil in the State vary widely from region to region. While it is fertile alluvial soil in the coastal plains and the narrow strips of river basins, it is of low fertility in the northern plateau. Mixed red and black soils are found in the Central Table land and in the ghats region.

The climate of the State is tropical in nature with high humidity. Average temperature ranges from a minimum of 12°C to a maximum of 39°C. The mean annual rainfall is 1482.2 mm with considerable variations in both space and time. About 85 per cent of the rainfall occurs during the period from June to September. Further although the mean annual rainfall in the State is 1482.2 mm, it is erratic over the years. As a result, if there is flood in some year, it is drought in another. Even these two opposite phenomena occur in the same year in two different regions of the State.

Population

The population of the State as per 1991 census is 3,15,12,070 of which 1,59,79,904 are males and 1,55,32,166 are females. Urban and rural population are 42,32,455 and 2,72,79,615 respectively. Density of population as per 1991 census is 202. There is concentration of population in the coastal belt; highest concentration being in Cuttack district with density figure of about 505 persons per sq. km. On the other hand the lowest density of about 77 persons per sq km. occurs in Phulbani district.

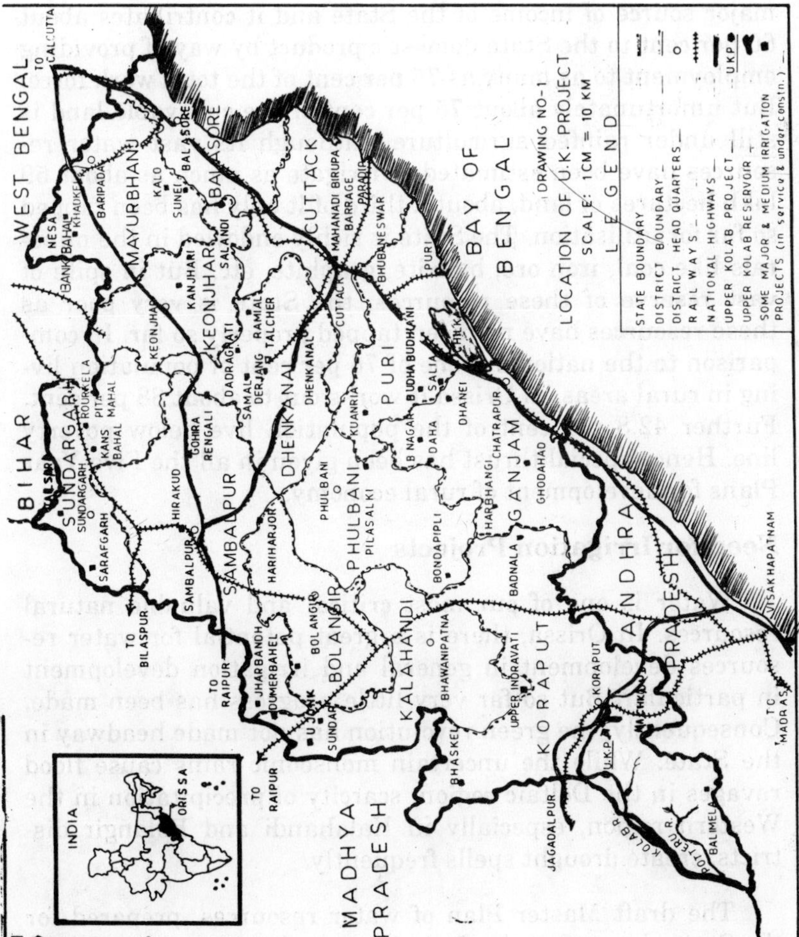
DRAWING NO-1
LOCATION OF U.K. PROJECT
SCALE: 1CM = 10 KM
LEGEND
STATE BOUNDARY
DISTRICT BOUNDARY
DISTRICT HEAD QUARTERS
RAILWAYS
NATIONAL HIGHWAYS
UPPER KOLAB PROJECT
UPPER KOLAB RESERVOIR
SOME MAJOR & MEDIUM IRRIGATION PROJECTS (in service & under constn.)
UKP
INDIA
ORISSA
N
S
BIHAR
WEST BENGAL
MADHYA PRADESH
ANDHRA PRADESH
BAY OF BENGAL
SUNDARGARH
MAYURBHANJ
BALASORE
KEONJHAR
CUTTACK
DHENKANAL
SAMBALPUR
BOLANGIR
PHULBANI
PURI
GANJAM
KALAHANDI
KORAPUT
TO CALCUTTA
TO BILASPUR
TO RAIPUR
TO MADRAS
VISAKHAPATNAM
JAGADALPUR
KOLAB R.
POTTERU R.
BALIMELA
MACHKUND
KORAPUT
U.K.P.
HIRAKUD
MAHANADI BIRUPA BARRAGE
PARADIP
BHUBANESWAR
CHILKA
ROURKELA
BARIPADA
UPPER INDRAVATI

Economy

Agriculture is the mainstay of the people of Orissa. It is a major source of income of the State and it contributes about 60 per cent to the State domestic product by way of providing employment to as many as 75 per cent of the total work force. But unfortunately about 75 per cent of the cultivable land is still under rainfed agriculture. Although its vast water resources have been estimated to irrigate as much as about 59 lakh hectares of land, about a third of it only has been tapped so far for utilisation. The State is richly endowed in the minerals like coal, iron ore, bauxite, graphite, etc. But in spite of vast reserve of these resources, the State is very poor as these resources have not been tapped properly so far. In comparison to the national scene of 76 per cent of population living in rural areas, in Orissa it works out to about 88 per cent. Further 42.8 per cent of the population live below poverty line. Hence special thrust has been given in all the Five-Year Plans for development of rural economy.

Need for Irrigation Projects

Water is one of our most critical and valuable natural resources. In Orissa, there is a great potential for water resources development in general and irrigation development in particular. But so far very little progress has been made. Consequently, the green revolution has not made headway in the State. While the uncertain monsoonic rains cause flood ravages in the Deltaic region, scarcity of precipitation in the Western region, especially in Kalahandi and Bolangir districts, create drought spells frequently.

The draft Master Plan of water resources, prepared for the State shows that the State has got about 4 per cent of the country's water resources in reserve. But planning comprehensive strategies for all round development of water resources had not been possible until 1951-52, when planned development in the country was initiated. At that time there were only two major irrigation system in the State; namely the Orissa canals and the Rusikulya system. The irrigation potential created through these two systems and few other small pre-plan projects was only 1.76 lakh hectares.

About thirty-five years of irrigation development till the commencement of Seventh Plan period, had raised the net irrigated area to about 17.22 lakh hectares which is about 26.255 of the net sown area and 29, 19 per cent of the feasible irrigable area. One of the objectives of the Seventh Plan was to bring as much cultivable area as possible under irrigation, particulary in tribal, economically backward and drought prone areas. In order to achieve this objectives, the programme for the Seventh Plan was to create additional 4.32 lakh hectares of irrigation potential. But due to several factors, both avoidable and unavoidable, the target could not be achieved.

More thrust has been given on Agriculture during the 8th Plan period. Irrigation being the basic input of agriculture, the latter's growth is very much dependent upon the former. Hence it is very much necessary to ensure that irrigation development is commensurate with agricultural development. But in spite of all efforts, progress of irrigation development in the State has not picked up the national average. According to the report of the Planning Commission—Working Group for Major and Medium irrigation Projects during seventh Plan, while many States will achieve their full potential by the turn of the century (*e.g.*, Manipur in 1997, Gujarat and Rajasthan in 1998, Jammu & Kashmir in 1989), in the case of Orissa, it is estimated that full potential will be achieved in 2055 A. D. Other agriculturally developed States like Andhra Prades, Karnataka, Tamilnadu and Punjab will achieve it in 2003, 2012, 2002 and 2002 respectively.

Thus, the need of irrigation project cannot be over emphasised. A close study of the planned development would reveal that about 50 per cent of the State's outlay has been in the field of irrigation and power. Again, a half of it is for irrigation sector. While it was Rs. 375.30 crores only during the plan period from 1951-52 to 1979-80, the investment during the two plan periods from 1980-81 to 1989-90 is Rs. 1105.00 crores. The outlay for Eighth Plan period is Rs. 2614.33 crores for Major and Medium Irrigation Sector alone. It has been the accepted policy to maximise irrigation potential creation to make the benefit reach the water scarcity pockets, especially in tribal and economically backward areas.

For effective and optimum utilisation of the country's vast water resources, a "National Water Policy" has also been formulated in 1987. The objective is to develop the available water resources for various beneficial uses like irrigation, hydropower generation, drinking water, industrial development and for other purposes. There are two different schools of thought; for and against large reservoir projects. On account of several adverse impacts on the physical and social enviroment due to construction of large reservoir projects, while one school of thought discourages construction of the said projects, the other school of thought favours them because of their multi-necessity character. It is no gainsaying that small dams are no substitute for large dams. Both are necessary. With proper management, the adverse effects of large dams can be minimised and the advantages can be promoted. Further, irrigation projects especially reservoir projects being mostly location specific, a judicious decision should be taken before going in for a large reservoir or a small reservoir project.

3

Irrigation Projects Involving Displacement of Human Settlement

3

Irrigation Projects Involving Displacement of Human Settlement

Classification of Projects

Irrigation is mainly classified into two types *i.e.,* flow and lift. Depending on ayacut capacity, flow irrigation is again classified into three types *i.e.,* major, medium and minor. A Major Irrigation Project is capable of irrigating more than 10,000 hectares of land. A Medium Irrigation Project is one, which can irrigate more than 2000 hectares upto 10,000 hectares. A Minor Irrigation Project irrigates less than 2000 hectares. On the other hand, the lift irrigation project has got no such yardstick. But on an average, it irrigates about 25 to 30 hectares of land.

Disturbance of Environment

Construction of flow irrigation projects involves impounding of water, resulting thereby in disturbance of environment. There is no such disturbance in case of lift irrigation projects.

Displacement of Human Settlement

Impounding of water may be caused due to construction of either a dam or a diversion weir/barrage. Further impounding causes submergence of land and property including human settlement. The more extensive the water spread area of the reservoir/pond is, the greater becomes the area of submergence. Normally the diversion weir/barrage project and the minor irrigation projects cause either no displacement or less displacement of human settlement in comparison to large and medium irrigation projects. The study here, therefore, is confined to the large and medium irrigation projects in Orissa.

Table 3.1 is the list of completed and ongoing Major, Multipurpose and Medium Irrigation Projects of Orissa and the magnitude of human displacement due to their construction.

The Table will indicate that as many as 56,377 of families comprising about 2.7 lakh people have been or in the process of being displaced from their ancestral home ground. In other words displacement of as big population as one third the population of Phulbani district is involved due to cons- truction of all these projects. The process does not end here. The Master Plan prepared for the irrigation development of the State shows that additional as many as 37 major projects and 62 medium projects can be constructed. The magnitude of human displacement on account of construction of these future projects has not been assessed. But the tentative assessement in the year 1983 for the proposed Manibhadra Multipurpose Dam Project alone indicates that it will affect 77,500 people in 233 villages in Cuttack, Puri, Phulbani and Dhenkanal districts.

Rehabilitation

What has been done so far for the displaced persons? Keeping the rehabilitation policy of the State Government in view, rehabilitation arrangements have been made in respect of some of the major and medium irrigation projects as indicated in Table '3.2'. It would indicate that rehabilitation by way of cash grant in lieu of land has been given more weightage.

The problems arising out of displacement and the manner in which they have been tackled especially in respect of Rengali Dam Project have been debated in several forums including the State Legislative Assembly. We shall discuss them in brief later at appropriate place.

Table 3.1

Human Displacement from Multipurpose, Major and Medium Projects in Orissa

Sl. No	*Name of project*	*Number of displaced families* *General*	*S.C.*	*S.T.*	*Total*
Multipurpose					
1.	Hirakud (completed)	N.A.	N.A.	N.A.	22,144
2.	Balimela (completed)	N.A.	N.A.	N.A.	1,200
3.	Salandi (completed)	32	5	352	569
4.	Rengali (completed)	8,015	1,710	1,172	10,897
5.	Upper Indravati (ongoing)	1,557	338	1,630	3,725 (not final)
6.	Upper Kolab	1,308	442	1,421	3,171
	Total	10,912	2,495	4,575	41,706
Major Irrigation Projects					
1.	Subarnarekha (Ongoing)	2,246	416	6,382	9,044
2.	Rengali (Ongoing)	N.A.	N.A.	N.A.	735
	Total	2,246	416	6,382	9,779
Medium Irrigation Projects					
1.	Dadraghati (completed)	228	66	133	427
2.	Derjang (completed)	327	29	-	356
3.	Baghua (completed)	406	8	96	510
4.	Ghodahad (completed)	3	-	65	68
5.	Baghlati (ongoing)	17	15	91	123
6.	Dumerbahal (completed)	74	23	156	253
7.	Pilasalki (ongoing)	57	24	104	185
8.	Kuanria (completed)	49	54	48	151
9.	daha (completed)	3	10	7	20
10.	Remal (completed)	1	2	1	4
11.	Sarafgarh (completed)	26	3	0	29
12.	Jharbandh (completed)	120	2	4	126
13.	Talsara (completed)	1	3	19	23
14.	Gohira (completed)	51	18	74	143
15.	Ramiala (ongoing)	166	159	89	414
16.	Sunet (ongoing)	84	2	267	353
17.	Kanjhari (ongoing)	113	4	80	197
18.	Bankbahal (ongoing)	98	35	149	282
19.	Kansbahal (ongoing)	10	29	172	211
20.	Hariharjore (ongoing)	140	41	213	394
21.	Harbhangi (ongoing)	17	8	128	153
22.	Badnala (ongoing)	4	2	163	169
23.	Upper Jonk (ongoing)	46	30	225	301
	Total	2,041	567	2,284	4,892
	Grand Total	15,199	3,748	13,241	56,377

(*Source*-Records of State Irrigation Department)

Table 3.2
Rehabilitation Arrangements made in some of the Multipurpose, Major and Medium Irrgtation Projects in Orissa

Sl. No	*Name of the Project.*	*Total No. of famil ies dis-placed.*	*No. of fa-milies re-settled in colonies.*	*No. of famili-es given cash grant.*	*Remarks*
1.	Rengali Dam	10,897	2,986	7,911	-
2.	Upper Kolab	3,171	528*	2,643	*104 moved to colonies in 1991
3.	Upper Indravati	3725(not final)	-	3,725	Rest will shift in phases.
4.	Dumerbahal	253	11	242	-
5.	Pilasalki	185	59	126	-
6.	Kuanria	151	40	111	-
7.	Daha	20	20	nil	-
8.	Remal	4	nil	4	-
9.	Sarafgarh	29	nil	29	-
10.	Jharbandh	126	nil	126	-
11.	Talsara	23	22	1	-
12.	Gohira	143	129	14	-
13.	Ramiala	414	383	31	-
14.	Sunei	353	169	--	Balance not decided.
15.	Kanjhari	197	47	125	25 to be resettled.
16.	Bankbahal	282	nil	282	-
17.	Kansabhal	211	-	-	shifting not started
18.	Hariharjore	394	171	223	-
19.	Harbhangi	153	-	-	shifting not started
20.	Badnala	169	-	-	"
21.	Upper Jonk	301	-	132	Balance not shifted so far.
Total		21,201	4,565	15,725	

(*Source*-Records of State Irrigation Department)

Large and Multi-purpose Projects, not only displace large population, but also disturb the environment in a bigger scale. The long gestation period also results in cost over run. Hence there are many advocates who support construction of medium and minor irrigation projects in preference to large and multipurpose projects. But the latter have their own inherent plus points which cannot be had from the former. It is not the forum to discuss these issues here.

4

Policy Parameters of Human Resettlement

4

Policy Parameters of Human Resettlement

Policy of Union Government

The preceding chapter gives a picture about the scene prevailing in Orissa. With a view to exploiting the water resources potential for the benefit of the people, a large number of major and multipurpose irrigation projects have been constructed in the country since launching of the First-Five Year-Plan in 1951-52.

There has not been any comprehensive documentation of the displacement of population due to construction of all these projects. However, the following Table 4.1 gives some idea about the quantum of displacement in some of the major projects in the country.

Displacement in all these projects and other similar projects is mostly due to submergence of the area inhabited by people. On account of geographical location of most of our projects, the problem arising out of displacement, touches mostly the weakest sections of the population like the S.T. and S.C.

Instructions of Government of India, 1980 (Ministry of Energy and Irrigation)

Unfortunately, so far there has been no national policy on resettlement of displaced persons. Hence individual States adopted their own rehabilitation policy. Each Ministry has taken care of the rehabilitation requirements as and when the occasion arose. Secretary, Irrigaton in the erstwhile

Table 4.1 : Quantum of Displacement in Some of the Major Projects in the Country

S.No.	Project	River	State	No.of villages affected. Fully	No. of Villages affected partly	No.of families affected.	Population affected.
1.	Sardar Sarovar	Narmada	M.P.	-	193	14.994	45,000
			Maharashtra	-	36	1.655	11,000
			Gujarat	3	16	4.700	10,593
			Total	3+245=248		21.349	66,593
2.	Narmadasagar	Narmada	M.P	254		N.A.	80,576
3.	Subarnarekha	Subarnarekha	Bihar&	182		9,722	50,116
			Orissa	24(fully)	40	9,044	N.A.
		(partly)					
4.	Ghataprabha	Ghataprabha (tributary of Krishna)	Karnataka	22		N.A.	15,660
5.	Malaprabha	Malaprabha (tributary of Krishna)	"	43		"	41,000
6.	Upper Krishna	Krishna	"	169		16,017	N.A.
7.	Nagarjunsagar	"	Andhra Pradesh	57		5,100	25,500
8.	Srisailam	Krishna	Andhra Pradesh	78		22,000	73,460
9.	Tehri Dam	Bhagirathi	Uttar Pradesh	95 +1 town		N.A.	46,000
10.	Tungabhadra	Tungabhadra	Karnataka	90		12,000	55,000
11.	Hirakud	Mahanadi	Orissa/M.P	249		22,144	N.A.

(*Source*-Records of State Irrigation Department)

Ministry of Energy and Irrigation, Department of Irrigation, Government. of India in his letter No.27(9)/80-P:I, dt. 19.5.80 to all State Governments requested to ensure protection to the weakest sections of the society while framing projects in the State. It aimed at creating in the oustees an interest in the execution of the project and one way of accomplishing this was to assign Government lands and waste lands in the Command Areas to them. In the event of non-availability of Government lands or waste land, it may be worthwhile to acquire land in the Command area from the larger land holders on the ground that their land would get water and productivity would increase. It was also stipulated in the said letter that "unless satisfactory safeguard are kept protecting the interests of the oustees particularly the weaker sections, Government of India might not entertain acceptance of the Project since rehabilitation issue might hold up the progress of the projects and result in excessive cost escalation. The paradox is that these warnings have remained as such, without being translated into action.

Instructions of Government of India (Ministry of Industry) 1986.

Thus, "land for land" was the only objective of human resettlement. But there was no clearcut guidelines in achieving this objective. Government of India, in the Ministry of Industry, Department of Public Enterprises, Bureau of Public Enterprises in their Office Memorandum No.15/13/84 B.P.E(C) dt. 3.2.1986 issued a set of instructions on land acquisition and rehabilitation aspects involved in major projects and alleviation of the difficulties faced by the dispossessed persons. Ministry of Water Resources in their letter No.6/71/84-P.II dt. 27.5.1986 impressed upon all State Governments to follow the instructions contained in the aforesaid Office Memorandum of Ministry of Industry, Department of Public Enterprises, while formulating projects. The salient features of the Office Memorandum are the following:-

(*a*) There should be a Rehabilitation Cell in each and every project to identify the persons who are dispossessed of their land following acquisition of their land.

(*b*) Rehabilitation assistance should be given to those who are dispossessed of their land and homestead and provided they were themselves cultivating those lands. Absentee landlords will not be entitled for any rehabilitation assistance.

(*c*) Persons losing whole of their land will get priority over persons losing land partially and persons losing homestead land will be compensated by allotment of developed homestead with all civic amenities like roads, water supply, educational, medical facilities, etc., cost of which will be borne by the project.

(*d*) Considering the educational attainment of the oustees, arrangements at project cost will be made to impart suitable-vocational training in available training institutes for employment in the project, subject to availability of vacancy. If suitable training disciplines are not available in existing institutes, such training courses which would equip the candidates for employment should be started and funded by the project. But there will be no commitment by the project authorities that such training arranged for by the project will ultimately provide employment to the oustee in the project itself or anywhere else. The idea behind such training is only to equip the oustee with necessary ability to compete for employment alongwith others.

(*e*) There will be no assurance for any employment opportunities in the project. But the projects will assist the State, Governments in organising and financing the oustees in taking to useful avocations like polutry, animal husbandry, etc. The basic responsibility of initiating such activities will be that of the State Governments; not the project.

(*f*) The Rehabilitation Cell will monitor the progress of rehabilitation activities.

(*g*) In case of tribal displaced persons, the Ministry of Welfare and Ministry of Environment and Forests should be associated with the preparation of their rehabilitation schemes.

(*h*) The entire cost of rehabilitation will be a part of the project cost.

Instructions of Govt. of India (Ministry of Home Affairs).

Under the Constitution, the tribal people in our country are entitled to receive several benefits of safeguards. Be it as it may, the major irrigation projects affect the tribals substantially. In this background, the Ministry of Home Affairs issued a set of guidelines on rehabilitation of displaced tribals following the recommendations of the conference of Chief Ministers held by the Union Home Ministry on 17.2.82.

For successful rehabilitation of displaced tribals, the following measures were suggested.

(*a*) Tribals should not be evicted from any area without making full payment of compensation for their land.

(*b*) Due to several enactments regarding sale of land belonging to tribals correct market value of land belonging to tribals is not known. Hence correct assessment of compensation amount should be done so that the tribal gets his due. Compensation amount should be deposited in Bank so that the tribal oustee gets some interest. If he intends to purchase land somewhere, he should be assisted in this regard by project authorities.

(*c*) Land based rehabilitation being the best form of rehabilitation, due care should be taken to allot land to tribals as far as practicable.

(*d*) Emergency procedure should not be adopted to acquire tribal land.

(*e*) As in the case of Maharashtra, law should be enacted in other States to allot land to the oustees in the Command Area of the irrigation project.

(*f*) Project report for resettlement of oustees should be prepared well in advance, keeping in view their socio-economic status and their absorption capacity.

(*g*) Displaced tribals should be rehabilitated not as individuals but as community as far as practicable.

(*h*) In case of non-availability of land, at least one member of the tribal family should be provided with employment. If need be they should be imparted training at project cost to acquire due proficiency to get employment either in the main project itself or in other departments of the State Government.

(*i*) Training programme of entrepreneurial skill, financial support, etc., should be arranged for self-employment.

(j) Free house sites should be provided to the affected families where houses of the oustees have been acquired.

(*k*) All civic amenities should be provided in the colonies set up for resettlement of the oustees.

Recommendations of Working Group

The Working Group on development of scheduled tribes during seventh Five-Year Plan in their report as published by the Government of India, in the Ministry of Home Affairs devoted one full Chapter on "Rehabilitation of Displaced Tribals."

The Working Group, *inter-alia* recommended the following:-

(*a*) There should be a national policy on rehabilitation of displaced persons with the following main items:-

(*i*) General principle of rehabilitation covering all categories of displaced persons.

(*ii*) special provisions regarding rehabilitation benefits to be extended under particular type of projects, and

(iii) special treatment to be given to tribal oustees.

(b) A project displacing tribals should give preference to tribals in jobs under it (atleast one member from each displaced family) as far as possible. For that matter training facilities should be extended to them well in advance at the project.

(c) In ancillary and finished goods to the main project (which displaced population) jobs for displaced tribals should also be reserved.

(d) Supporting services like dairy, poultry, fishery, carpentry, etc., around the main project should be encouraged exclusively through tribal people/tribal cooperatives etc.

(e) Group rehabilitation should be the rule in case of displaced tribals and all civic amenities should be provided there.

(f) Displaced triblas who are agriculturists, should be given preference to settlement in agriculture; and landless triblas should be settled in other avocations of their own like as far as possible.

(g) Land allotment for tribals as far as practicable should be made in the Command Area; or else irrigation facilities should be provided in areas other than Command Areas.

(h) There should be a fullfledged Directorate of Rehabilitation in every State and a State Level Rehabilitation Committee under the Chairmanship of Chief Secretary should be formed to review programme and progress of rehabilitation.

The Working Group also recommended for payment of compensation money towards land acquisition in terms of capitalised value of income as the sale statistics available in tribal area do not give the correct picture on account of protective legislations in the area.

Standing Committee for Rehabilitation of Tribals

Although the Eighth Five-Year Plan has already commenced, no national policy has been formulated so far, excepting constitution of a Standing Committee for review of rehabilitation of tribals displaced by major projects, vide Government of India, Ministry of Welfare Office Memorandum No. 20017/3/86 TDB dt. 29.8.1986. It is a national level committee head by the Union Deputy Minister (Welfare) as Chairman constituted keeping the guidelines of the Bureau of Public Enterprises, Ministry of Industry in view. The Committee is required to monitor the efforts of the project authorities and the State Governments in implementing thier rehabilitation schemes and to ensure speedier rehabilitation of the displaced tribals. Later the group of ministers on Tribal unrest in its meeting dt. 26.3.87 considered and approved an Approach Paper on "National Policy on Rehabilitation of Displaced Tribals." After detailed delibrations, the committee of Secretaries of Government of India processed a draft "National Policy on Rehabilitation of Tribals displaced under the Major Projects". It awaits approval of Government of India.

National Water Policy

Thus, as the matter stands, there are no specific guidelines for the resettlement and rehabilitation of persons displaced due to construction of irrigation projects; even though a "National Water Policy" has already been framed in September 1987. The National Water Policy, among others prescribes that "there should be an integrated and multi disciplinary approach to the planning, formulation, clearance and implementation of projects, including catchment treatment, enviromental and ecological aspects, the rehabilitation of affected people and Command Area Development." But this in itself is not sufficient to settle down the issue.

As a sequel to this, Government of India, in the Ministry of Water Resources, in their order No. 17/4/90-PP dt. 15.3.91 constituted a Sub-Committee of the National Water Board under the Chairmanship of Secretary, Ministry of Welfare to review the various rehabilitation policies already notified by the State Governments, to prepare guidelines for resettlement of project affected people and to prepare a draft National Rahabilitation Policy. The Policy is still awaited.

Regional Seminars on Environmental Issues-1987

The Water and Power Consultancy Services (WAPCOS), a Government of India Undertaking under the Ministry of Water Resources conducted four Seminars on "Water Resources Development—Enviromental Issues and Solutions," between September 1987 and November 1987 on regional basis at Bhopal, Ahmedabad, Trivandrum and Bhubaneswar. Rehabilitation and resettlement of oustees of irrigation projects was one of the themes discussed in these Seminars. Subsequently a national seminar was held at New Delhi on 29.2.1988. All these Sminars *inter alia*/made the following main recommendations

(*a*) A national policy for the resettlement and rehabilitation of oustees is formulated and implemented effectively.

(*b*) In view of the special socio-economic situations of tribal people, it should be ensured that these people as far as possible are resettled as viable communities.

(*c*) Pending formulation of a National Policy the Central Government may issue suitable guidelines immediately based on the decisions taken in a number of projects which may serve as a common frame for resettlement and development of the people already affected by the projects in different States.

Thus, there is need for a national rehabilitation policy for the irrigation project displaced persons.

Orissa Rehabilitation Policy—No Uniform Policy

As there is no national rehabilitation policy, there is no uniform rehabilitation policy for the development projects executed by the State Government. This is the reason for which different norms are followed for the rehabilitation of oustees from the "National Test Range—Balipal", "National Thermal Power Station—Talcher" and even the State managed Ib Thermal Power Station. However of late, in a high level meeting dt. 8.6.90, presided over by the Hon'ble Chief Minister, Orissa, some policy decisions were taken to have a uniform rehabilitation policy relating to displacement of oustees of various types of development projects. This was followed by Irrigation Department Resolution No. 31888 dt. 21.8.90. The salient freatures have been discussed in succeeding paragraphs. Our subject of study being resettlement and rahabilitation of oustees from irrigation projects, let us limit our discussion to the irrigation projects only.

Hirakud Dam Project

Historically the problem of rehabilitation assumed importance in Orissa only when the construction of Hirakud Dam was taken up just before plan development through Five-Year Plans commenced in the country. The seriousness of the problem can be seen by the fact that as many as 22144 families in 249 villages in Orissa, besides affecting 36 villages in Madhya Pradesh were displaced. The vast artifical reservoir of about 743 sq km. submerged as much as 1,23,303 acres of cultivable land. As per the policy followed then, displaced persons were paid full compensation for lands and properties acquired from them. They were free to resettle themselves in places of their choice or opt to resettle in Government sponsored colonies. In the latter case, reclaimed lands were allotted to them at a subsidised rate of Rs. 213 per acre. Cost of reclamation was Rs. 500 per acre, and it was borne by the project. There was no limit to the extent of land for allotment. House sites were given free in the reclaimed areas and timber was supplied at 60 per cent of the normal royalty for the construction of houses. Manures and fertilisers were supplied free of cost and seed paddy was given on loan. "Katas" and "Mundas" for irrigation purposes, tanks and wells for drinking water purposes, school buildings, hospital, community centre, "bhagabat ghar" etc. were provided in the settlement colonies. Transport was free of cost to the submerged area people for carrying all their movabls excepting straw.

The Hirakud Land Organisation was incharge of the responsibility of orderly removal of the families from the reservoir area to their places of choice for resettlement. It was not an easy task. The complexity of the problem can be understood by the fact that even now complaints are received from some of the displaced persons about payment of compensation money.

Balimela and Rengali

Hirakud was followed by Balimela in the early seventies. It displaced about 1200 families and the policy adopted in Hirkud was almost repeated in Balimela. But Rengali saw a new chapter in the history of irrigation development in Orissa. Initially local people did not welcome the construction of Rengali Dam Project; the obvious anger being against the manner in which problems of displacement were being tackled. There was great agitation which consequently delayed the completion of the project. But this agitation paved the way for the Government to approach the displacement and resettlement problems squarely in a systematic manner. Thus, was born a new and uniform Rehabilitation Policy.

Orissa Rehabilitation Policy

As a result of the experience in Hirakud and Balimela, a cogent policy on resettlement and rehabilitation of oustees of Rengali Dam Project was drawn up in 1973 by the Orissa Government. This policy has also been made applicable to all the major and medium irrigation projects in the State. The policy is an improvement on the policy adopted for Hirkud and Balimela and the two main improvements are, (a) lands for resettlement in Government sponsored colonies are allotted to the oustees free of salami irrespective of the fact whether the oustee was a landless or not, and (b) displaced persons who desire to make their own rehabilitation arrangements in lieu of land, are given cash grant, equivalent to the cost of the land due to him.

The salient features of the policy as were in force then and made applicable to Upper Kolab Project are discussed below:-

Displaced Person

A person is considered displaced who on account of acquisition of land for the purpose of major or medium irrigation project has been displaced from such lands including any landless and homeless person who is dependent for his livelihood, by manual labour, on agricultural lands immediately before the area comprising such lands is taken up for aquisition by the project and who is being deprived of such livelihood on account of acquisition of the lands.

Displaced Family

Family in relation to a displaced person means such person, and his or her spouse, minor sons, unmarried daughters, minor brothers or sisters, father and mother and other members residing with him and dependent on him for their livelihood. (In 1989 it was amended that major sons aged 21 years and above and who are married will be trated as separate families. In 1990, the age limit was further relaxed to 18 years irrespective of any marital status).

Enumeration

The date of laying the foundation stone of the project, *i.e.*, 11.6.75 was the reference date for counting of displaced families. Later in 1987, the date of publication of notification under section 4(1) of the Land Acquisition Act, 1984 was taken as the date for the purpose of enumeration of families.

Eligibility

The folllowing principle is adopted to determine eligibility in the matter of allotment of land:-

(*a*) In fully submerged villages, displaced families losing land and/ or houses will get land in rehabilitation area provided they do not have three acres or more land in any adjoining area.

(*b*) Landless and homeless families belonging to the fully submerged villages will also get land.

(*c*) In partially submerged villages, the following categories of displaced families will get land-

(i) Displaced families retaining houses but losing their land, thereby suffering a reduction in their holding to less than three acres in the same or adjoining villages.

(ii) Displaced families losing houses and lands and thereby suffering a reduction in holding to less than three acres in the same village or adjoining village.

(iii) Displaced families initially having holdings of less than three acres in the same village or adjoining village but losing house, irrespective of the fact whether they lose their agricultural land or not.

(iv) Displaced families retaining houses but losing encroached land in the same village or adjoining village so as to render them landless.

(v) Landless and homeless families.

Facilities

(1) *Land for Land* has been the objective of resettlement. Over and above the payment of compensation money for the land and houses, trees, etc., acquired for the projects, land as under was allotted free of salami to each oustee family subject to availability.

(*a*) A 0.5 of homestead land for habitation.

(*b*) 6 acres of reclaimed unirrigated land or 3 acres of reclaimed irrigated land for cultivation purposes.

In 1989, it was amended to provide 5 acres of reclaimed unirrigated land or 2.5 acres of reclaimed irrigated land. It was further amended to 2.5 acres of unirrigated land or 1.25 acres of irrigated land and homestead plot of 20 decimals.

In case of non-availability of required extent of land, allotment is reduced, Initially the cost of reclamation of land is borne by the project, subject to a maximum of Rs. 960 per acre. 30 per cent of the reclamation cost *i.e.,* Rs. 300 per acre is recovered at a later stage from the allottee as a first charge from the compensation money payable to the person concerned. Subsequently in 1989 the reclamation cost and recovery amount was enhanced to Rs. 1050 and Rs. 315 per acre respectively. In the event of non-recovery, due to some reason or other, the said amount is recoverable as arrear of land revenue. However, full recovery of the compensation money is recovered from oustees who receive compensation of Rs. 50,000 and above and no recovery is made from the oustee who are landless.

(2) Construction of Houses

Displaced families are allowed free transport by the project authorities for shifting as well as for carrying their house building materials which they salvage from old sites to the new settlement colony. They are also provided with house building forest material at concessional rate 60 per cent of normal royalty.

(3) Provision of Common Facilities

common civic amenities like schools, drinking water wells and tanks, club houses, etc., are provided in the settlement colonies at the following scale:-

(*a*) *(i)* Up to 44 families...one well

(ii) For 45 to 70 families...Two wells

(iii) For 75 to 100 families...three wells

In case of individual resettlement in existing Revenue villages, where the number of displaced families between 10 and 14 settle down, one drinking water well is also provided.

(*b*) One tank for each colony irrespective of number of families (*c*) One two-roomed school building for colonies with less than 50 families and a well within its campus.

(*d*) One club house irrespective of the number of families and a well near it.

(*e*) internal roads in settlement colonies.

(4) Miscellaneous Facilities

Besides facilities are provided for reconsecration of deities. For amenities like education for higher secondary and higher education, public health centre, veterinary dispensary, Panchayat and Panchayat Samiti Offices, respective Administrative Departments of Government, supplement the provisions made in the Rehabilitation estimates from their Departments budget. Electrification of the colonies is made as far as practiable. Care should be taken to allot the agricultural lands as far as practicable in the irrigation command areas; otherwise facilities for minor irrigation should be provided as far as possible.

(5) Payment of Cash in Lieu of Land

Wherever land is not available for resettlement of the displaced families or when the displaced families wish to make their own rehabilitation arrangements, rehabilitation grant of Rs. 14,040 at the flat rate of Rs. 2160 per acre (Rs. 1200 towards land +Rs. 960 towards cost of reclamation) was given in lieu of the land (6 acres of unirrigated land+1/2 acre of homestead land) to be allotted to them. This grant is in addition to the compensation money given to the persons towards acquisition of land and houses etc. (This ceiling of Rs. 14,040 was enhaced to Rs. 20,075 in 1989 and further enhanced in 1990).

(6) Employment

Over and above the aforesaid facilities Government also provide employment opportunities. 50 per cent of the vacancies in unskilled posts in the project will be kept reserved for the people of the submersible areas.

(7) Vocations

In an ideal set up, with a view to encouraging the oustees to take to useful avocationas like poultry, pisciculture, horticulture, animal husbandry, etc.

Government will assist in organising and financing such activities. Training should be imparted at project cost to equip the eligible oustees to be considered for employment in the project. Of course, such training imparted should not be presumed to be a commitment for ultimate employment in any organisation.

Administration

Spl. L.A. and R.O

Although the Chief Executive of the Project is in overall charge, for land acquisition and resettlement work, Revenue Officers are posted exclusively. In case of major and multipurpose project he is designated as Special Land Acquisition and Rehabilitation Officer and in respect of medium irrigation project, he is designated as Land Acquisition and Resettlement Officer. While the former works in the project itself, his salary, etc., being charged to the project, the latter works under the direct control of the respective Collectors. Keeping the volume of work in view, the post of Spl. Land Acquisition Officer is separated from that of Resettlement Officer. In other words two independent officers are detailed; one for Land Acquisition and the other for Resettlement. If need be, a third officer is posted as in the case of Rengali Multipurpose Project, designated as Deputy commissioner, Rehabilitation to coordinate the work between the Spl. L.A.O. an Resettlement Officer.

Advisory Committee

The policy also provides for constitution of Rehabilitation Advisory Committee consisting of officials and non-officials including peoples' representatives of the area affected and benefited. In respect of medium irrigation projects, such committees are headed by the respective Collectors of the district and in respect of major and multipurpose projects and those projects which spread over more than one district; the committees are headed by the respective Revenue Divisional Commissioner. Government decides to appoint any one of the Revenue divisional Commissioners as the Chairman of the Committee. Government may also constitute High Level Committees under the Chairmanship of the Chief Minister to resolve rehabilitation aspects. Such a Committee is there in respect of Rengali Multipurpose Project.

Rehabilitation Policy Relating to Other Projects in the State

We will discuss later about the extent, manner and propriety of implementation of this policy in respect of oustees of Upper Kolab Project. But prior to it, we would consider it desirable to throw some light on the policy parameters adopted in some major projects other than irrigation, now under execution in the State. We will also discuss in brief about guidelines formulated by World Bank on involuntary resettlement and the policies adopted in other States in respect of oustees of irrigation projects there.

This discussion we consider necessary to have a balanced view about the policy being adopted in different sectors *vis-a-vis* the status prevailing in Upper Kolab Project on which the case study is being done.

Talcher Super Thermal Power Project

National Thermal Power Corporation have been entrusted with the execution of this project at the present cost of Rs.1480.85 crores. On completion it will have 1000 M.W. installed capacity and 600 M.W. firm capacity.

The total requirement of land for the project is 4120 acres. Due to execution of this project, 1750 *No. of* families will be affected of whom 1092 families will be substantially affected and balance 658 families will be partially affected. National Thermal Power Corporation have prepared a comprehensive Rehabilitation Scheme in consultation with the State Government; and the same is under consideration of the Government of India.

The socio-economic survey conducted to obtain various data revealed that general choice of the would be displaced persons was for employment. It also revealed that there was inadequate availability of cultivable waste land. Therefore, the alternative rehabilitation options available are:-

(a) Training in I.T.I.

State Government will provide technical and vocational training in existing institutes in the State, particularly in trade like Welders, Electrician, Operators, etc., for which job opportunities exist in the power plant. For areas where no job opportunity exists in the plant, the oustees after training can avail of job opportunities in other institutions and industries etc. The training cost will be met by N.T.P.C. but induction into training institutions willl be done by State Government. This opportunity will be opened on priority to substantially affected persons.

(b) Allotment of Shops Adjacent to Plant Approaches and Inside Township

110 persons in the first category and 50 persons in the second category can be provided with this benefit. Building for shops will be constructed by the project and will be allotted on long term basis to those persons who do not have aptitude/qualification for undergoing training. Besides the compensation money received towards acquisition of land etc., seed money shall be arranged as financial assistance from nationalised banks.

(c) Engagement in Agro-based Industries

65 families who were originally marginal farmers will be given a plot of land of 0.5 acre for growing vegetable and another 0.10 acre for homestead and shed for keeping either dairy/poultry. The shed will be constructed for them.

(d) Unskilled Job in the Plant

About 50 people will be engaged in development and maintenance of horticulture both at plant site and township on regular basis. 100 people can get permanent job in sweeping and cleaning work in the plant as well as in the township.

Thus, 1092 number of families who are substantially affected (losing more than 1/3rd of their land) will be engaged in avenues like training, shop allotment and agrobased industries. The balance 658 families will have lot of opportunities for their income in construction, execution and other fields like horticulture, service contracts and small industrial enterprises as self-employment or direct employment. They will continue to reside in their existing homesteads.

(e) Homestead Oustees

50 families who would lose their homestead land or more than 1/3rd of landholdings will be given developed house site alongwith basic infrastructure. But the oustees will construct their houses at their own expense. The plant authorities will provide free transportation for shifting of salvages to their new place of settlement.

Ib Thermal Power Station, Banaharpalli

This project aims at generating 840 M.W. (4 x 210 M.W) power utilising the low grade coal deposits in the Ib Valley in Sambalpur district at a total estimated cost of Rs.1200 crores.

Construction of the project will displace 178 families from three villages namely Banharpali, Telenpali and Bangad.

Keeping ,the guidelines of Revenue and Excise Department in view, the Orissa Power Generation Corporation which is constructing the project, has formulated a comprehensive rehabilitation policy for the oustees of this project.

The guidelines of Revenue and Excise Department as referred to above were issued in Notification No.GE (SBP)96/ 88-71 dt. 2.1.89 in connection with rehabilitation of displaced persons due to coal mining operation in the districts of Sambalpur and Dhenkanal by the South-Eastern Coalfields Ltd. Keeping the said guidelines in view O.P.G.C. have formulated a Rehabilitation Scheme for the oustees. Its salient features are:-

(a) Eligibility

A family/person losing all the lands, or losing more than 1/3rd agricultural land alongwith homestead, or losing only homestead land will be eligible for rehabilitation benefits. Homesteadless persons are also eligible.

(b) Family

Family will mean husband and wife as the case may be, their unmarried children, but does not include a major son who has been separated from the family on or before the date of notification under section 4(1) of the L.A.Act, 1894. He will be treated as a separate family.

(c) Employment

Employment opportunities to one member from each of the following categories of families will be provided by the O.P.G.C. in order of priority:-

(*i*) Families losing dwelling houses, homestead lands and not less than 1/3rd of total holding.

(*ii*) Families losing 3 acres of unirrigated land or 2 acres of irrigated land.

Employment as under will also be provided subject to availability.

(*I*) Families losing only homestead land or the total agricultural holding.

(*II*) Families losing 1/3rd of the total agricultural holding.

Where the oustee does not have any adult member in the family who is suitable for employment his son-in-law or adopted son will be eligible for employment.

Rehabilitation of other families shall be made through self-employment schemes. For this O.P.G.C. have constructed shops and pedestals to be allotted.

(d) Rehabilitation Colony

Each displaced family will be allotted 10 decimals of reclaimed Government land for homestead purposes free of cost and all civic amenities will be provided in the colony. For construction of temporary shed in those colonies, each family is required to be given cash grant. In the instant case OPGC have programmed to construct a two-roomed house with Verandah, kitchen covering till the oustees make their own permanent arrangement.

(e) Transport

Each family will be given an allowance of Rs. 2000 for shifting their belongings from the area of evacuation to the settlement colony.

National Test Range, Baliapal

Altogether 11,609 families with a population of 70,102 are likely to be displaced.

As in the case of the previous two projectes, this project mainly aims at vocationally oriented rehabilitation as land is not available for allotment for agricultural purposes. However, in the safety zone identified by the project authorities, agricultural land acquired for the project will be leased out on temporary basis subject to the condition that no dwelling houses will be constructed.

The displaced persons will be resettled in model villages with the following facilities.

(*a*) 10 decimals of homestead land free of cost.

(*b*) Built up house on the homestead land costing about Rs.15,000. No cash will be given in lieu of the said house.

(*c*) All civic amenities including electricity, dispensary, Bank, Post Offices, Community Centres, etc.

(*d*) Some built up shops will be provided in each model village.

(*e*) There will be a hospital and a fire brigade for the settlement area.

(*f*) There will be several industries, and an industrial training institute.

(*g*) Large number of fishermen families to be displaced will be settled in one village also with all civic amentities.

Jobs in the Test Range and in the industries to come up will be offered to displaced persons on priority. The objective is to offer job opportunity to one member of each displaced family under direct as well as self-employment scheme. Direct employment will be available in the industries and in the Test Range; whereas self-employment opportunities will be given by way of introducing schemes for the landless fishermen and other suitable persons. Nets and boats will be provided to fishermen and the landless displaced will be given two heads of milch cows after proper selection.

Each family will also get a grant of Rs. 1500 for transportation of their belongings to their new houses.

Decision for a Uniform Policy for the State

It has been indicated above that in Irrigation Department Resolution No.31888 dt. 21.8.90, a uniform rehabilitation policy has been spelt out. It has been further indicated that the resolution will have retrospective effect from 1.6.90. However, displaced families, who have already been provided with rehabilitation assistance either in shape of land or in shape of cash prior to 1.6.90 will not be eligible to get the benefits of this resolution. The modifications made thereon from time to time brought out in the aforesaid resolution and several amendments are summed up in the following:

Family Enumeration

A son aged more than 18 years of age will be treated to be a separate family, irrespective of his marital status.

Physically infirm and mentally retarded persons irrespective of their age, will be considered as separate families.

Minor orphans who have lost both the parents and have nobody to fall back upon, will be considered as separate family. Such orphans numbering more than one in a particular family will be clubbed together and counted as one family.

Divorcees with dependent child/children and divorcees without children but having no source of livelihood will be considered as separate families.

Reference Date for Enumeration of Families

The date of publication of notification under section 4(1) of the L.A.Act, 1894 will continue to be the date of reference. However, the list of families to be entitled to rehabilitation benefits will be updated as on 1st January of the year in which the actual displacement takes place.

House Building Assistance

It was the rule that the oustees will make necessary arrangements to build their own houses as they liked after availing of houses building forest materials at concessional rate. But latest decision is either to provide each displaced family resettling in project sponsored colonies with a house of given specification at project cost or to pay a cash grant of Rs. 17,000 per family to construct a house as per the specifications approved by the Project authorities. Those families opting to stay outside the project sponsored colonies are also entitled to get Rs. 17,000 in cash as house building assistance.

There is no longer any provision to supply house building forest materials to the oustees.

Maintenance Allowance

Each and every displaced family irrespective of the fact that one resides in the project sponsored colony or at a place of his choice will get maintenance allowance of Rs.500 per month for one year provided he/she evacuates as per the programme framed by the project authorities.

Acquisition of Private Land

Private land in the Command Area of the irrigation projects may be acquired for allotment to the oustees. Necessary enactment in this regard will be made after examining the pros and cons of laws of the land. All transfers of private land in the Command Area after issue of Notification U/S 4(1) of the L.A.Act may be banned too. Further purchase committees may be formed to assist the oustees in the purchase of private land.

As regards Government non-forest land, in the Command Area and nearby areas may be earmarked for rehabilitation purposes. 'Gochar' lands in the Command Area may also be converted into cultivable Government land outside the Command Area.

Allotment of Government Land

Each displaced family will be provided with:

(*a*) Homestead plot of 20 decimals.

(*b*) Agricultural land.

Each displaced family including landless family will get either 2.5 acres of unirrigated land or 1.25 acres of irrigated land for agricultural purposes.

(*c*) The status of land, whether irrigated or unirrigated at the time of allotment will be taken into consideration, without taking into account whether there is future scope of irrigation facilities or not. However, land to be acquired, if any, in the Command Area for allotment to oustees, will be treated as irrigated land.

(*d*) A person who loses land through acquisition for construction of canal system and who is entitled to get compensation for such acquisition will get an equal area of land if unirrigated or half the area of irrigated but not more than 2.5 acres of unirrigated land or 1.25 acres of irrigated land.

A person losing his only house due to acquisition for canal construction and who is entitled to get compensation for such acquisition will get an equal area of house site subject to maximum of 20 decimals and house building assistance as admissible to displaced persons of submersible area.

(*e*) A person of the affected village, who has no landed property whatsoever, will not be entitled to any rehabilitation benefits, if the village is not evacuated due to construction of canal.

Rehabilitation Cash Assistance

In case the project authorities are unable to allot land for rehabilitation purposes or the displaced person opts in writing to make his own arrangements, rehabilitation cash assistance in lieu of land will be given to the displaced families at the rate of Rs. 8000 per acre of unirrigated land or Rs.16,000 per acre of irrigated and homestead land, subject to the ceiling outlined above. This amount is inclusive of the cost of reclamation. Further it will be released after negotiation for purchase of private land and at the time of its actual registration or for self-employment schemes with or without Bank assistance.

Employment

(*a*) 50 per cent of the unskilled and clerical posts in the hydro projects will be filled up from among the oustees of the project and the balance will be filled up by the employees found surplus in other completed projects.

(*b*) As far as possible all unskilled and clerical posts in other industrial and commercial projects will be filled up from among the oustees of the project.

Organisation

There will be a separate Directorate of Rehabilitation under the Revenue Department which will be responsible for the rehabilitation of the oustees of all projects in the State. The Directorate will be multidisciplinary in composition. Suitable officers will be brought on deputation from other Departments to deal with the entire gamut of rehabilitation activities.

Participation of Voluntary Organisations

Execution of certain components of the rehabilitation plan may be entrusted to reputed voluntary organisations.

Thus, as the modifications decided upon indicate, the approach of Government to rehabilitation of project oustees is very liberal. But the author is tempted to comment that.

(*a*) No option should be left with project authorities to provide built up house in the project sponsored colonies and thereby make displaced families dependent upon Government machinery. Instead of providing built up houses, the oustees could be provided with all facilities to acquire house building materials and technical know-how to build their own low cost houses. Under "Indira Awas Yojana," a good number of residential colonies have been set up by Government at different places for the weaker section of people. But instances are not rare, where these colonies now stand deserted. This unhappy experience should not be lost sight of while providing ready built houses to the oustees in resettlement colonies.

(*b*) Further by reserving 50 per cent of the unskilled and clerical posts for the oustees in the hydro projects and all such posts in other projects, Government have deprived other eligible candidates of the area who are not displaced, from competing for these posts. As per current practice, thousands of N.M.R. (Nominal Muster Roll) employees are there in innumerable time bound construction projects. They all are required to be retrenched immediately their term is over or the project is completed. The problem becomes compounded when these retrenched employees remain unemployed. Further reservation provisions under rules of employment should not be lost sight of.

(*c*) Physically infirm and mentally retarded persons irrespective of their age will be counted as separate families. It has to be ensured that there is no misutilisation of this benefit and the interest of this category of families is not disturbed by anybody. Such persons should not be enrolled as beneficiaries under any other poverty eradication schemes.

(*d*) It is good that 'divorcees' who have no source of livelihood are treated as separate families for rehabilitation purposes. But depending solely on affidavit and enquiry report of the officers of R and R organisation may offend the legal provision under Divorce Act.

(*e*) The existing revenue laws prescrible a minimum percentage of "gochar" land and land for community purposes. Conversion of gochar land into cultivable land may geopardise the interest of that particular village community. They will face difficulties in community the distance from that village to the gochar land that will be made available elsewhere.

(*f*) It is indeed heartening that those oustees who would opt to make their own rehabilitation arrangements as per their choice would get house building assistance and the amount will be paid in three instalments. But since these oustees will leave the submerged area to places, as they like, it will be a difficult job to monitor their rehabilitation and release the instalments of aforesaid house building assistance each time after verification of proper utilisation of money. Hence there is every apprehension that the oustees may fritter away the house building cash assistance as they do in respect of compensation money and rehabilitation grant.

It may be mentioned here that, the Rehabilitation policy adopted by Orissa is granted to be the most beneficial policy as it aims at rehabilitating the oustees by way of allotment of land. But in practice it has been seen that, due to some reason or other, rehabilitation grants in cash is being given more and more in lieu of land and the amount has been enhanced substantially from Rs. 2160 to Rs. 8000 per acre of unirrigated land during the period of last one year. This enhancement will prompt the oustees more and more to avail of cash money in preference to allotment of land on the pretext that they will make their own rehabilitation arrangements. Further, provision for payment of house building assistance to those oustees who will settle down outside the project sponsored colonies by making their own arrangements may defeat the very purpose of systematic and successful resettlement of oustees. There is apprehension that oustees, in their expectation to get cash money will be tempted to avoid to move to the project sponsored colonies.

Another important factor which has been lost sight of is that there is no distinction between Project Affected person and Displaced person for the purpose of rehabilitation. Not necessarily a person affected due to construction of the project will be displaced from his place of residence.

Policies Adopted in Other States

All these discussions lead us to the conclusion that while the policy for the oustees of irrigation projects is based on allotment of land, the policy in respect of projects other than irrigation is employment and vocation oriented. Agriculture being the prime occupation of about 80 per cent of our people, it has been argued that allotment of land to the oustees will help in rehabilitating them in a more meaningful way. That is why the Rehabilitation Policy in respect of oustees of irrigation projects in Orissa is considered to be the most liberal policy in the country. In this context it is necessary to know about the policies adopted in other States. We have taken few selective States in this regard and their policies are discussed in brief in a statement form to give a comparative picture vide Table 4.2.

Table 4.2
Comparative Statement Showing the Rehabilitation Policy of Orissa and Other States

Orissa	*Andhra Pradesh*	*Bihar*	*Gujarat*	*Karnataka*	*Madhya Pradesh*	*Maharashtra*	*Tamilnadu*	*Uttar Pradesh*
1	*2*	*3*	*4*	*5*	*6*	*7*	*8*	*9*
(A) Displaced Person								
Any person who on account of acquisition of his land for the purpose of major & Medium Irrigation Projects has been displaced from such lands including any landless and homeless person, who is dependent for his livelihood by manual labour on agricultural lands immediately before the area comprising such lands is taken up for acquisition by the project and who is being deprived of such livelihood on account of acquisition of the land.	There is difference between a person displaced and a person affected. The affected persons are those who are actually involved in shifting and hence do not include absentee land owners or house owners but will include their tenants.	Those persons who lose their residential houses, due to submergence or who suffer economically due to damage to land & property are enumerated as displaced persons. Those persons, who after acquisition of land to the extent of half or more are left with 2.5 acres or less of irrigated land or 3.2 acres or less of unirrigated land are also treated as oustees.	Any person residing in submergence area, cultivating land, carrying on trade or getting livelihood by any other means one year prior to the notification under section 4(1) of L.A. Act.	Displaced person means any tenure holder,tenant, Govt.lessee or owners of other property who on account of acquisition of his land including plot or other property in the affected zone as declared by Government is displaced from such land and other property.	Displaced person means any tenure holders, tenants, Govt. lessee or owners of other property and whose land including abadi plots have been displaced from such land or property.	Any occupant who due to acquisition of his land in the affected zone of the project is displaced from such land or any agricultural labourer is treated as displaced person.	-	-
(B) Family (i) Family in relation to a displaced person	Family means a person and his dependents i.e. son, dau-	Husband & wife, minor sons and unmarried daughters only con-	Family shall mean husband, wife, minor children and	Family means the displaced person and his or her spouse,	Family consists of displaced person and his or	Family of the displaced person means such person and his		

(Contd.)

Table 4.2 (Contd.)
Comparative Statement Showing the Rehabilitation Policy of Orissa and Other States

Orissa	Andhra Pradesh	Bihar	Gujarat	Karnataka	Madhya Pradesh	Maharashtra	Tamilnadu	Uttar Pradesh
1	2	3	4	5	6	7	8	9
means such person, and his or her spouse, minor sons, unmarried daughters, minor brothers and mother, minor sisters, father and other members residing with him and dependent on him for their livelihood. *(ii) Major sons* who are 18 years of age and are married or not. *(iii) Physically* infirm and mentally retarded persons irrespective of their age. *(iv) Minor* orphans who have lost both the parents and have now body to fall back upon.	ghter, spouse.	stitute a family. Any adult above 18 years of age is treated as a separate family.	persons dependent on head of the family. Every major married son shall be considered as separate family.	minor sons, unmarried daughters, minor brothers or sisters, father and mother & other members residing with him/her and dependent on him/her for their livelihood.	her spouse, minor son, unmarried daughter, minor brother, or unmarried sister, father and mother and other members residing with him and dependent on him for their livelihood.	or her spouse, minor sons, unmarried daughters, minor brothers or sisters, father & mother and other members residing with and dependent upon him/her for their livelihood.		

(Contd.)

Table 4.2 (Contd.)
Comparative Statement Showing the Rehabilitation Policy of Orissa and Other States

Orissa	*Andhra Pradesh*	*Bihar*	*Gujarat*	*Karnataka*	*Madhya Pradesh*	*Maharashtra*	*Tamilnadu*	*Uttar Pradesh*
1	*2*	*3*	*4*	*5*	*6*	*7*	*8*	*9*
Such orphans numbering more than one in a family will be clubbed together and counted as one family. **(v) Divorcees** with dependent child/children and divorcees without children but having no source of livelihood. 2. *Enumeration* The date of publication of notification U/S 4(1) of L.A. Act, 1894 is reference date for the purpose of enumeration of families. List of families will be updated on 1st January of the								

(Contd.)

Table 4.2 (Contd.)
Comparative Statement Showing the Rehabilitation Policy of Orissa and Other States

Orissa	*Andhra Pradesh*	*Bihar*	*Gujarat*	*Karnataka*	*Madhya Pradesh*	*Maharashtra*	*Tamilnadu*	*Uttar Pradesh*
1	*2*	*3*	*4*	*5*	*6*	*7*	*8*	*9*
year in which actual displa--cement takes place.								
3. *Facilities provided to the displaced persons.*								
(A) Land (Agricultural)								
Each displaced family including landless family, will get either 2.5 acres of un-irrigated land or 1.25 acres of irrigated land. A person losing land through acquisition for ment of construction of canal system and who is en-titled to get	In case of submersion of tribal lands, rehabi-litation will be on land to land basis.	A displaced family dependent on agri-cultre shall get priority in allot-ment of at least 2 acres of land for agricultural purposes. S.T. & S.C. ousteees will get special weightage in allotment of land. Every landless, small and marginal family will get	The affected family whose land is acquired due to submerge-nce, will be allotted irri-gable agricul-tural land equal to what was acquired from him/her with a minimum of 2 Ha and maximum limited to land ceiling in the State. Every major son of the\ landed oustee, every joint holder and his/	Allotment of land free of cost to tena-nts and agri-cultural labo-urers in a graduated scale depend-ing on area of land lost by the dis-placed person, minimum being one acre & maximum being four acres of irriga-ble agri-cultural land and 2 acres in the mi-	Land equal to that is acqui-red subject to a minimum of 2 hectares and maximum of 8 hectares to all land hol-ders and mini-mum of one ha. and maximum of 2 Ha is allotted to encroachers who encroached the land in affected area prior to April, 1987. Landless persons are not eligible	Every land holder ous-tee will be eligible to get land eq-uivalent to the land ac-quired, sub-ject to pre-scribed ceiling and a minimum of 2 Ha. Every joint holder will get equal to his share going under	Priority will be given for allotment of cultiva-ble waste lands to the persons who have been dispossessed of their lands under direct cultivation whether as an owner or as a tenant due to ac-quistion	Those opt-ing for land are given min-imum of 2 acres in lieu of their holding on pay- cost which is deducted from compensa-tion money. Landless agri-cultural lab-ourer families get 2 acres free of cost or Rs.40,000/-

(Contd.)

Table 4.2 (Contd.)
Comparative Statement Showing the Rehabilitation Policy of Orissa and Other States

Orissa	*Andhra Pradesh*	*Bihar*	*Gujarat*	*Karnataka*	*Madhya Pradesh*	*Maharashtra*	*Tamilnadu*	*Uttar Pradesh*
1	*2*	*3*	*4*	*5*	*6*	*7*	*8*	*9*
compensation for such acquisition will also get an equal area of land subject to a maximum of 2.5 acres/1.25 acres of unirrigated & irrigataed land respectively.		a grant of Rs.10,000/- for purchase of land and Rs.15,000/- for its development. All possible help will also be given to them under normal development programmes.	her major son, every landless agricultural labourer of the submergence area and each of his major son and encroschers on non-forest Govt. land and even those encroaching forest land in the area going under submergence and each of their major sons will also be eligible for 2 hectres of land for agriculture. If Govt. land is not available, project authroities will arrange for purchase	nimum and 8 acres in the maximum of non-irrigable agricultural land. Non-irrigable non-agricultural land of 5 acres in the minimum is granted to displaced persons holding 0.6 acres of land before acquition or an agricultural labourer.	to get land but they will be provided with other means of economic rehabilitation. Ceiling surplus land in command area will be acquired and made available to oustees on payment. 50% of the sum of compensation money will be set apart towards cost of land. Balance cost of the land allotted will be recovered in 20 yearly instalments. The affected	submergence subject to minimum of 2 hectares. Every encroacher whose encroachment is prior to 31.3.78 to get 2 Ha. Landless acricultural labourer to get one hectare if he moves out alongwith other oustees.	for public purpose resulting in reduction of their holding to below 3 acres of wet land provided that the land value shall be collected if compensation had been paid for the land acquired.	cash compensation.

(Contd.)

Table 4.2 (Contd.)
Comparative Statement Showing the Rehabilitation Policy of Orissa and Other States

Orissa	*Andhra Pradesh*	*Bihar*	*Gujarat*	*Karnataka*	*Madhya Pradesh*	*Maharashtra*	*Tamilnadu*	*Uttar Pradesh*
1	2	3	4	5	6	7	8	9
			of priyate land through Purchase Committee headed by the Addl. Collector, assisted by local M.Ps, M.L.As, Panchyat Presidents, representatives of N.G.Os. Representative of PAPs also remain present in identification of land. In case land identified is higher in price than the compensation money received by the oustee the difference is entirely borne by the Govt. and is treated as 100% subsidy.		families will also be encouraged to purchase land directly.			
(B.) Land (Homestead)								
Each displace family will be allotted	The affected family belonging to weaker	Each displaced family will	Each PAP family is allotted a resi-	In case members of a displaced	The oustee agriculturist families will	Every oustee family, its major		Development house sites are allotted for constucting houses and shops.

(Contd.)

Table 4.2 (Contd.)
Comparative Statement Showing the Rehabilitation Policy of Orissa and Other States

Orissa	*Andhra Pradesh*	*Bihar*	*Gujarat*	*Karnataka*	*Madhya Pradesh*	*Mahara-shtra*	*Tamil-nadu*	*Uttar Pradesh*
1	*2*	*3*	*4*	*5*	*6*	*7*	*8*	*9*
0.20 acres of land free of cost in the resettlement colonies to be set up by Government. A person losing his only house due to acquisition for canal construction and who is entitled to get compensation for such acquisition will get an equal area of homestead land subject to a maximum of 0.20 acres.	section who are eligible for al' tment of ho se sites/ house under the ongoing "sites and service" programme will be allotted house sites and also given hutment grant of Rs.1000/- per family. Where Govt. lands are available the affected families shall be allotted house sites and where ever it is necessary private lands have to be acquired for	be allotted 0.25 acres of land free of cost. Those who want only 0.125 acre will get the value of rest 0.125 acre land. Priority will be given to oustees in allotment of houses under Housing Board Schemes.	dential plot measuring 9ft X60 ft in close proximity to his agricultural land.	family do not exceed 5, a plot measuring 1200 sq.ft to 3994. 73 sq.ft is given and where the members exceed 5, a plot measuring 2400 sq. ft to 5992. 13 sq.ft is given.	be allotted a 150 sq.mt. residential plot and the other oustee families will get 100 sq.mt plot free of cost.	son and that of land holder oustees and encroachers shall be eligible to get 5400 sq.ft. of house plot.		

(Contd.)

Table 4.2 (Contd.)
Comparative Statement Showing the Rehabilitation Policy of Orissa and Other States

Orissa	Andhra Pradesh	Bihar	Gujarat	Karnataka	Madhya Pradesh	Mahara-shtra	Tamil-nadu	Uttar Pradesh
1	2	3	4	5	6	7	8	9
	the purpose. Affected families other than weaker sections will be allotted house sites to the extentt possible on payment of market value.							
(C) Civic amenities to be provided in resettlement colonies.								
1- Wells. Upto 44 families—one 45 to 70 families—Two 70 to 100 families-Three. In case of individual resettlement in existing Revenue villages where the number of displaced families is at least 10- one well will be provided.	No colonies are set up by the Government. Displaced persons are only given ex-gratia cash grant in lieu of rehabilitation at given rate depending on the nature of acqusition of their land (vide E below).	One well for every 50 fami-lies and one tube well for every 50 families.	One well and one Havada shall be givven at new settlement site and one addl. well shall be given per 100 families.	One well for every 15 to 20 families.	One drinking water well with trough for every 50 families	As in case of Madhya Pradesh.		All civic amenities are provided in settlement colonies.

(Contd.)

Table 4.2 (Contd.)
Comparative Statement Showing the Rehabilitation Policy of Orissa and Other States

Orissa	*Andhra Pradesh*	*Bihar*	*Gujarat*	*Karnataka*	*Madhya Pradesh*	*Maharashtra*	*Tamilnadu*	*Uttar Pradesh*
1	*2*	*3*	*4*	*5*	*6*	*7*	*8*	*9*
(ii) Tank, One tank for each colony irrespective of number of families.		One tank for every rehabilitation colony.	percolation tanks or village tanks shall be given at the new places of resettlement wherever possible. Arrangements should be made so as to give the benefit of tanks to the groups of oustees.	-	One tank for every 500 families.	As in case of Madhya Pradesh.	-	-
(iii) School One two roomed school building for colonies with less than 50 families and a well within the campus.	Schools are to be provided at Project costs.	One Primary school for every rehabilitation colony.	A school shall be given for 100 families at the new places of resettlement.	Primary schoool of required dimension.	One primary school for every 100 families. Middle School if existed previously.	As in case of Madhya Pradesh.	-	-
(iv) Club, Panchyat Ghar etc. One Club House irrespective of the number of families and a well near it.		One Panchyat Bhavan, one play Ground, Childern Park for every rehabilitation colony.	Where the total number of families of the adjoining resettlement sites become 500, a Panchyat or Samaj Bhavan is to be provided.	one village Panchyat or Community Hall.	One Panchayat Ghar-cum-Community Hall for every 500 families.	One Panchyat Ghar for every 500 families.	-	-
(v)Internal roads in resettlement colonies.	Internal and connecting roads shall be provided.	Approach road to link the main road and internal colony roads.	Internal roads as well as road connection to main roads are to be provided.	Approach roads as well internal colony roads.	Road to connect main road.	As in case of Madhya Pradesh.	-	-

(Contd.)

Table 4.2 (Contd.)
Comparative Statement Showing the Rehabilitation Policy of Orissa and Other States

Orissa	*Andhra Pradesh*	*Bihar*	*Gujarat*	*Karnataka*	*Madhya Pradesh*	*Maharashtra*	*Tamilnadu*	*Uttar Pradesh*
1	2	3	4	5	6	7	8	9
(vi) Facili--ties to be provided for reconsecration of deities.	Temples & Places of archaeological importance in submersible area will be shifted to safer place by Archaeology & Museum Deptt.	One religious Centre for each rehabilitation colony as per Caste.	Religious structures shall be constructed on exchange basis.	Temple or places of worship in lieu of which existed in the submerged area.	Provision for religious places of wor--ship, play ground, park, Cremation & Burial Ground etc.	One Platform for every 50 families.	-	
(vii) Facilities for higher education, public health centre etc. are to be provided by the concerned Administrative Deptt.		One middle school and one High School for every 500 families. One P.H. Centre for every rehabilitation colony.	After resettlement if total families of adjoining resettlement sites become 500, a hospital shall be provided.	-	One dispansary for every 5000 families.	As in case of Madhya Pradesh.	-	
(D) Employment. 50% of the vacancies in unskilled and clerical posts in Hydro Projects to be filled up by	Not more than 50 % of the vacancies of the categories equivalent to Jr. Asst., Typists and Cadre be-	50% of the vacancies in Class-III posts in the Project will be filled up from amongst the oustees.	Employment of oustees in the Project will be given as far as possible in following priority order.		The affected families of the submerged area will be given priority in employment in the project. Special efforts will be made to give employment to the educated youths.	To employ eligible candidates in various posts in the project activities and to	Members of the family including S.C. & S.T. whose lands have been acquired for the Govt.	Preference is given to displaced persons in services in the project.

(Contd.)

Table 4.2 (Contd.)
Comparative Statement Showing the Rehabilitation Policy of Orissa and Other States

Orissa	*Andhra Pradesh*	*Bihar*	*Gujarat*	*Karnataka*	*Madhya Pradesh*	*Maharashtra*	*Tamilnadu*	*Uttar Pradesh*
1	*2*	*3*	*4*	*5*	*6*	*7*	*8*	*9*
displaced persons. In other projects (Industrial) all unskilled and clerical posts will be filled up by the oustees.	low arising in Major & Medium.Irr.& Power Projects shall be filled up by the diplaced families or their dependents of a respective project duly following the reservation for various categories viz. S.C., S.T., Ex-Servicemen etc. The regular appointments made under this scheme shall be kept outside the purview of the A.P. Public Service	Preference will be given to oustees in all appintments to Class-IV posts. In all earth work, digging of wells etc. oustees will get priority. To open up avenues of self employment for families who were not dependent on agriculture training facilities at Govt. cost will be provided to atleast one member of the displaced family and will be given a sti-	(i) Oustees whose lands are fully sumberged. (ii) Oustees whose lands are partially sub--merged, oustees of S.C.,S.T. and economically backward community will be given top priority. Vocational training facilities are also provided at project cost.		If necessary, they will be given training also. The affected families will be engaged in new trade and to of business. Ancillary industries will also spring up as a result of construction work. The affected families will be given preference in allotment of shop and land. In order to promote the spirit of self employment amongst oustees, they will be adjusted in getting loans from the Commercial Banks to purchase Trucks, Tractors and Trollies etc. and the Project will help them in all respects and as far as possible, these machines will be utilised on the project work.	arrange for training in vocational trades..	purposes as well as for the project of the Public Sector Undertaking, subject to the condition that preference should be given to those who are dependent for their livelihood primarily or wholly on the land acquired and from among them to members of the S.C. & S.T. who may also be eligible for employment.	

(Contd.)

Table 4.2 (Contd.)
Comparative Statement Showing the Rehabilitation Policy of Orissa and Other States

Orissa	*Andhra Pradesh*	*Bihar*	*Gujarat*	*Karnataka*	*Madhya Pradesh*	*Maharashtra*	*Tamilnadu*	*Uttar Pradesh*
1	*2*	*3*	*4*	*5*	*6*	*7*	*8*	*9*
	Commission.	pend of Rs.5000/- per person. A loan of R.25,000/- on concessional rate of interest will also be given to start any income generating business.						
(E)Rehabilitation Cash Assistance.								
In case Govt. are unable to allot land or where somebody opts to make own resettlement, arrangements, rehabilition cash assistance well be given in lieu of land Rs.8000/-per acre of unirrigated land and Rs.16000 per acre of irrigated and homestead land subject to ceiling in 3.A.	Exgratia cash grants at the rate of 500% of the compensation amount is paid in lieu of rehabilitation subject to a maximum of Rs.1000/-where either lands or houses are acquired and Rs.5000/- where land as well as houses are acquird	-	-	-	Rehabilitation grant of Rs.6400/- for landless agricultural labourers, S.T,S.C., Small & Marginal farmers and Rs.3200/- to the landed oustees is given.	Resettlement grant of Rs.750/- with Jan'80 as the base & a rise of 8% per year thereafter.		
(F) Other Benefits								
(i) Free transportation	(i) Trahsporation and	(i) Every displaced	(i) Free transport for	Providing	(i) The project has the respon-	(i) Free transport		(i) Displacement grant for shift-

(Contd.)

Table 4.2 (Contd.)
Comparative Statement Showing the Rehabilitation Policy of Orissa and Other States

Orissa	*Andhra Pradesh*	*Bihar*	*Gujarat*	*Karnataka*	*Madhya Pradesh*	*Maharashtra*	*Tamilnadu*	*Uttar Pradesh*
1	*2*	*3*	*4*	*5*	*6*	*7*	*8*	*9*
of belongings from the old house site to new resettlement colonies. (ii) *Each* displaced family will get a house of given specification in the project sponsored colony. But the oustee is also at liberty to make his/her house in the project sponsored colony or outside and for which house building assistance @ Rs. 17,000/- per family will be given in three instalments.	shifting of the affected families to the nearest safe places by the project authorities. (ii) Opening of additional depot of Forest Deptt. to provide building materials to the displaced families. (iii) If displaced persons form themselves into Cooperatives, refinancing facilities for house building shall be arranged by Dist.Collector from the A.P. State Cooperative Housing Societies.	family who is entitled to get compensation up to a maximum of Rs.5000/- towards acquisition of land will get grant in aid of Rs.1500/- to Rs.5000/- in a graduated scale. This assistance also includes transport charges. (ii) Each displaced famely will get financial assistance of Rs. 20,000/- for construction of 2 roomed houses on the home-	shifting dismantled building materials. (ii) Exemption from Stamp duty subenable the oustees to purchase land if land is purchased within three years of award of his land acquired. (iii) Oustees are given building materials like Cement, iron Sheets etc. from the Irrigation Department depot at controlled price. (iv)Each oustees family will be paid a subsistence allowance at Rs. 15/- per day for 25 days in a month for one year from the actual date of residing in a new relocation site and continuously residing at that place for a	free transport from Place of mergence to new settlement site selected.	sibility to transport the personal effects of the affected families. The project has to arrange for trucks and to bear the expenses. (ii) Full exemption of registration fees and stamp duty to the affected families who purchase land in the command area. (iii) Exemption under the Ceiling Act to the affected families.	of personal kit, household articles and salvaged building material. (ii) Grant-in aid upto Rs.500/- as laid down in N.W.D.T. award.		ing household effects and merchandise of shops etc. is given memberwise in the family. For tenants living in rented houses or doing business in rented shops rehabilitation grant is given (indi-viduals @ Rs.500/-).. and families @(Rs.1000/-). Those oustees prefering land are paid rehabilitation grant @ Rs.1000/- per family for purchase of seed, manure etc. (ii) Construction materials like Cement, G.C. are supplied

(Contd.)

Table 4.2 (Contd.)
Comparative Statement Showing the Rehabilitation Policy of Orissa and Other States

Orissa	*Andhra Pradesh*	*Bihar*	*Gujarat*	*Karnataka*	*Madhya Pradesh*	*Maharashtra*	*Tamil-nadu*	*Uttar Pradesh*
1	*2*	*3*	*4*	*5*	*6*	*7*	*8*	*9*
		stead land attoted to him.	Period of complete one year.					From Project on cost.
(iii) Every oustee family whether residing in colony or outside will get maintenance allowance of Rs. 500/- per month for one year.			(v) Outright subsidy upto Rs. 5000/- to PAP family is given to purchase income generating assets like bullock carts. (vi) Insurance cover of Rs. 6000/- to oustee family against accident and death.					
(G) *Participation of Voluntary Organisation.* Execution of certain components of the rehabilitating plan mey be entrusted to reputed voluntary organisations.			Participation of NGOs is accorded highest priority			As in case of Gujarat		

Policy Guidelines in World Bank Financed Projects

The World Bank have come up in a big way to finance a good number of development projects in the developing countries. Presently in Orissa, they are financing Rs. 821 crores Upper Indravati Hydro-power Project. Until March '88 they were also funding 18 medium irrigation projects styled as Orissa Irrigation--II. Appraisal for the Subarnareka Major Irrigation Project is under progress. Reimbursement of cost of resettlement and rehabilitation is not admissible under the terms of agreement with the Bank. But the Bank considers it as one of the basic and mandatory fulfilments for meriting reimbursement on other items of work. It is, therefore, considered desirable to discuss in brief about the Bank's guidelines for World Bank aided projects in the matter of involuntary resettlement as brought out in "World Bank Technical Paper Number 80" by Michael M. Cernea.

(*a*) Displacement of human settlement should be avoided as far as possible, if unavoidable, displacement should be reduced to the minimum.

(*b*) All resettlement programmes must be development programmes as well so that the productive base and income earning ability of resettlers are improved and they share the benefits of the project which displaced them.

(*c*) Socio- economic survey of the affected area and population to be displaced should be conducted well in advance.

(*d*) It is the responsibility of the concerned Government to relocate the affected persons and for which an appropriate rehabilitation action plan with time schedule is prepared.

(*e*) While planning resettlement programmes as well as in executing them, affected people should be consulted either directly or through their representatives.

(*f*) To avoid misunderstanding, friction among the resettlers and the host habitats, it is necessary to meet equitable treatment in the matter of development activities, civic amenities, etc.

(*g*) The approach to resettlement programme should be to ensure that the oustees regain at least their previous standard of living and as far as practicable, they should be economically and socially integrated into the host communities. To achieve this, a realistic action plan should be prepared in a manner that would give the settlers the opportunity to become physically established and economically self-sustaining in the shortest possible time. The action plan should contain the following-

(*i*) Adequate compensation should be paid for the land and property lost by the oustee.

(*ii*) For the oustees losing agricultural land, agricultural redevelopment package should be prepared in the new set up. For this adequate extent of land should be identified which should have been reclaimed and have irrigation facilities.

(*iii*) Where land available is not suitable for agriculture, programmes should be made for tree and orchard development, social forestry, pisciculture, etc.

(*iv*) For landless, artisans especially when land is not available, non-land based programmes need to be framed. They include service sector, industrial sector, self-employment avenues, etc.

(*v*) Cash payment alone for resettlement and rehabilitation should be avoided. Alternative homes, employment opportunities, etc., should also be provided. "Land for land" approach should be rigidly followed.

(*vi*) There should be an Environment Management Plan so as to protect environment.

(*vii*) Oustees should be shifted to resettlement sites sufficiently ahead of impounding of reservoir.

(*h*) There should be distinct organisation to oversee the implementation of the rehabilitation and resettlement action plan backed by adequate budget provision.

(*i*) There should be timely monitoring of the progress and completion of resettlement activities, and the general social and economic condition of resettled people.

Thus, the primary objectives of the resettlement of oustees as laid down by the World Bank are (a) preparation of the affected people for the transfer from the submersible area (b) an orderly evacuation of the displaced people to the new site within a stipulated period and (c) to co-operate with and advise the displaced persons during the evacuation period and after, so that they readjust themselves in a manner whereby the social environment and economic opportunity of the re-settlers will be as desirable as that available to them prior to their evacuation, if not better.

As the matter stands, although there is no national policy on rehabilitation of oustees, different ministries and State Governments have issued guidelines to ensure timely and or-derly resettlement of them. We may now discuss about the situation prevailing in Upper Kolab Project. Before embark-ing upon to do so, it is necessary to know in brief the history of the project.

5

Upper Kolab Project

5

Upper Kolab Project

River

River Kolab, on which the Upper Kolab Project has been constructed rises near Sinkaram hill of the Eastern ghats in Koraput district of Orissa at an elevation of RL 1200 M. Its main tributaries are the Guradi and the Karandi. The major downstream tributary is Potteru. After traversing about 78 Km and collecting all the water available in the catchment area, it descends to a level of about 810 M at the Upper Kolab Dam site. A number of small rapids and falls were there before the river reaches the dam site. They all are now inside the reservoir. Lower down the dam there are innumerable rapids and falls too. Before it touches the Orissa-Madhya Pradesh border, it passes through a Zone of calcium rock bed. A finest specimen of "stalactites" and "stalamites" is seen in the 'Gupteswar' temple. The famous and massive 'Shiva Linga' in this rock-cave temple is a developed "stalagmite" and is worshiped by all Hindus. A great 'mela' is held here on 'Shiva Ratri' occasion.

The river forms common boundary between Orissa and Madhya Pradesh from village Godohati for some distance, then flows in Orissa territory and again forms the boundary of these two States till joins Sileru river at village Motu in Orissa. In the lower reaches, the river is called 'Saberi' and Motu is the trijunction point of Orissa, Madhya Pradesh and Andhra Pradesh. From this point the river flows as 'China Godavari' (Small Godavari) and falls into the 'Godavari' a few kilometres downstream of famous Bhadrachalam in Andhra Pradesh.

DRAWING NO-2
MAP OF
KOLAB CATCHMENT
SCALE: 1 MM = 1 KM
N
MADHYA PRADESH
ORISSA
KORAPUT
GURADI R.
KARANDI R.
KOLAB
MACHKUND
KOLAB
MACHKUND R.
BALIMELA
SAVERI
POTERU
SILERU R.
MOTU
ANDHRA PRADESH
R. GODAVARI
LEGEND
STATE BOUNDARY
RIVER & NALLA
CATCHMENT BOUNDARY
DAM SITE

Map 2

The river from its head upto Motu is 450 Km long and its catchment area intercepted at the dam site is 1630 sq. km. The course of the river and its complete catchment area are indicated in Drawing No. 2. The catchment area above the Upper Kolab Dam site is indicated in Drawing No. 3.

History of the Project

Preliminary investigations to utilise the water resources of Kolab river was started in 1961. Two alternative dam sites were investigated, a few metres downstream of the confluence of Guradi river with Kolab river, near village 'Panasput' (Lat 18° 47'N and 82° 37' E longitude). Both the dam sites were 218 metre apart from each other. Accordingly geological survey was taken up. But due to certain technical considerations, the dam site was shifted later by about 1.5 Km downstream near village 'Koranga' (Lat 18° 48N - Long 82° 37° E). Finally it was settled to be done at 'Koranga'. After detailed technical scrutiny in the Central Water Commission and Central Electricity Authority, the irrigation project was approved in May 1976 and the hydel project was approved in August 1975. Finally the administrative approval was accorded by the State Government in Irrigation and Power Department. L. No. 27966 dt. 5.8 1976 and No. 22806 dt. 24.6.1976 respectively. The infrastrcture development works were started after June 1976 and the dam construction was completed in 1986. The reservoir was impounded from 29.8.1986 and the first unit out of four units was commissioned on 18.12.1987. It was followed by 2nd Unit on 29.3 1988 and 3rd Unit on 10.2.1990. The 4th Unit was programmed to be commissioned in December 1991.

Upper Kolab Project being a multipurpose project, also aims at giving irrigation. Water discharged from the power house is conducted through tail race channel and caught by a small earthen dam across 'Satinallah' wherefrom the irrigation canals start. The intensity of irrigation programmed is over a C.C.A. of 47,710 hectares.

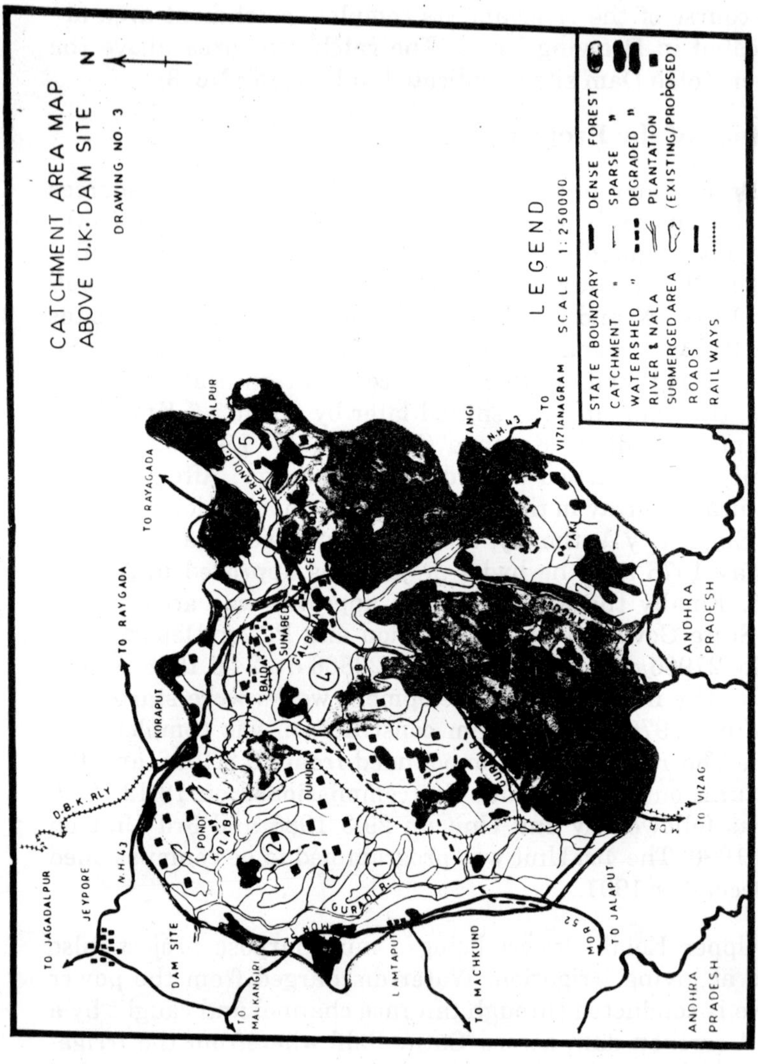
CATCHMENT AREA MAP
ABOVE U.K. DAM SITE
DRAWING NO. 3
N
LEGEND
SCALE 1:250000
STATE BOUNDARY
CATCHMENT "
WATERSHED "
RIVER & NALA
SUBMERGED AREA
ROADS
RAILWAYS
DENSE FOREST
SPARSE "
DEGRADED "
PLANTATION
(EXISTING/PROPOSED)
TO JAGADALPUR
JEYPORE
N.H. 43
D.B.K. RLY
KORAPUT
TO RAYGADA
TO RAYAGADA
DAM SITE
TO MALKANGIRI
TO MACHKUND
LAMTAPUT
TO JALAPUT
TO VIZAG
TO VIZIANAGRAM
ANDHRA PRADESH
PONDI
KOLAB R.
DUMURIA
GURADI R.
SUNABEDA
BALDA
SEMILIGUDA
KERANDI R.
PAKI
N.H. 43

Map 3

Profile of the Project

(a) Hydrology -

(i)	Catchment area	1630 sq. km
(ii)	Maximum weighted rainfall in catchment	1609 mm
(iii)	Minimum weighted rain fall in catchment	1165 mm
(iv)	Weighted average minimum rainfall in catchment	1415 mm
(v)	Maximum design flood discharge	7650 cumsecs
(vi)	Mean annual run off	180300 Ha. metre

(b) Dam

(i)	Type - straight masonry gravity type.	
(ii)	Total length	630.5 metre
(iii)	Length over non-over flow portion-	375 M
(iv)	Top of dam	862.5 M
(v)	Road width	7.5 M
(vi)	Deepest bed level	810.40 M
(vii)	Foundation level	808 M
(viii)	Maximum height of dam	54.50 M

(c) Spillway

(i)	Length including piers	255.50 M
(ii)	No. of spans	14
(iii)	Capacity of spillway	7650 cumsecs
(iv)	Type of gates	Radial
(v)	Size of gates	15 M x 7.5 M
(vi)	Crest level of gates	850.50 M

(d) Dykes - Two No. of dykes namely Landiguda and Jhatiguda.

The former is 257 M long and the latter is 466 M long.

(e) *Reservoir*

(i)	Full Reservoir Level	858 M
(ii)	Minimum draw down level	844 M
(iii)	Average reservoir level	853.33 M
(iv)	Gross storage capacity	1215 M. Cum
(v)	Storage capacity at minimum draw down level	280 M.Cum
(vi)	Live storage	935 M. Cum
(viii)	Regulated discharge	42.5 Cumsecs.

(f) *Tunnel-*

(i)	Type - Horse shoe type.	
(ii)	Length	3924 M
(iii)	Diameter	5.5 M
(iv)	Design Discharge	120 Cumsecs

(g) *Surge tank*

(i)	Type - Differential	
(ii)	Diametre	18 M
(iii)	Depth	108 M
(iv)	Maximum surge level	883.5 M
(v)	Minimum surge level	834.2 M

(h) *Penstock tunnel*

(i)	Type and Number - Horse shoe type - 3 Nos.	
(ii)	Length	972 M
(iii)	Diametre	4.75 M

(i) *Penstock*

(i)	Number	3 Nos.

(ii) Total length	940 M
(iii) Diametre	3 M
(iv) Thickness	50 mm

(j) Power

(i) Installed capacity	320 MW
(ii) Number of Units	(4 x 80 M.W)
(iii) Firm capacity	95 M.W.

(k) Satiguda pond

(i) Length of earth dam	942.5 M
(ii) Maximum height of dam	16.70 M
(iii) Maximum water level	595.20
(iv) Maximum pond level	587.68 M
(v) Spillway capacity	334 Cumsecs

(1) *Irrigation*

(i) G.C.A.	88,740 Hactares
(ii) C.C.A.-	47,710 Hectares
(iii) Right canal (Jeypore Main Canal)	
(A) Discharge at head	98.1 Cumecs
(B) Length	58.83 Km
(C) C.C.A	22267 Ha lift
	46049 Ha flow
(iv) Left canal (Padmapur distributory)	
(A) Length	12.6 Km
(B) C.C.A	1936 Ha flow
(C) Discharge at head	2.79 Cumsecs.

An outline map indicating different components of the Project is given in Drawing No.5.

Cost of the Project

The Project consists of three different units, *i.e. (a)* (Unit-1, Dam and appurtenant works, *(b)* Unit-II (Irrigation works) and *(c)* Unit-III - (Power Works). The cost of Unit I is apportioned between irrigation sector and power sector on 50:50

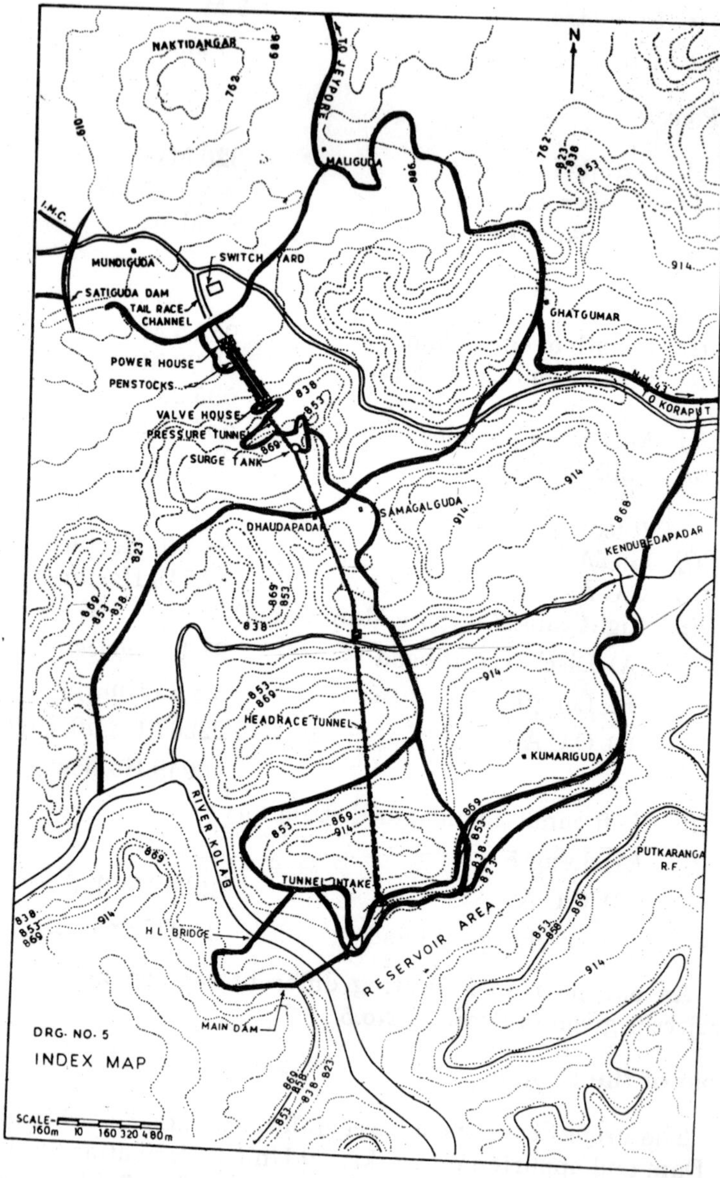
N
NAKTIDANGAR
TO JEYPORE
MALIGUDA
I.M.C.
MUNDIGUDA
SWITCH YARD
SATIGUDA DAM
TAIL RACE CHANNEL
GHATGUMAR
POWER HOUSE
PENSTOCKS
NH 43
TO KORAPUT
VALVE HOUSE
PRESSURE TUNNEL
SURGE TANK
SAMAGALGUDA
DHAUDAPADAR
KENDUBEDAPADAR
HEADRACE TUNNEL
KUMARIGUDA
RIVER KOLAB
PUTKARANGA R.F.
TUNNEL INTAKE
H L BRIDGE
RESERVOIR AREA
MAIN DAM
DRG. NO. 5
INDEX MAP
SCALE
160m 10 160 320 480m

Map

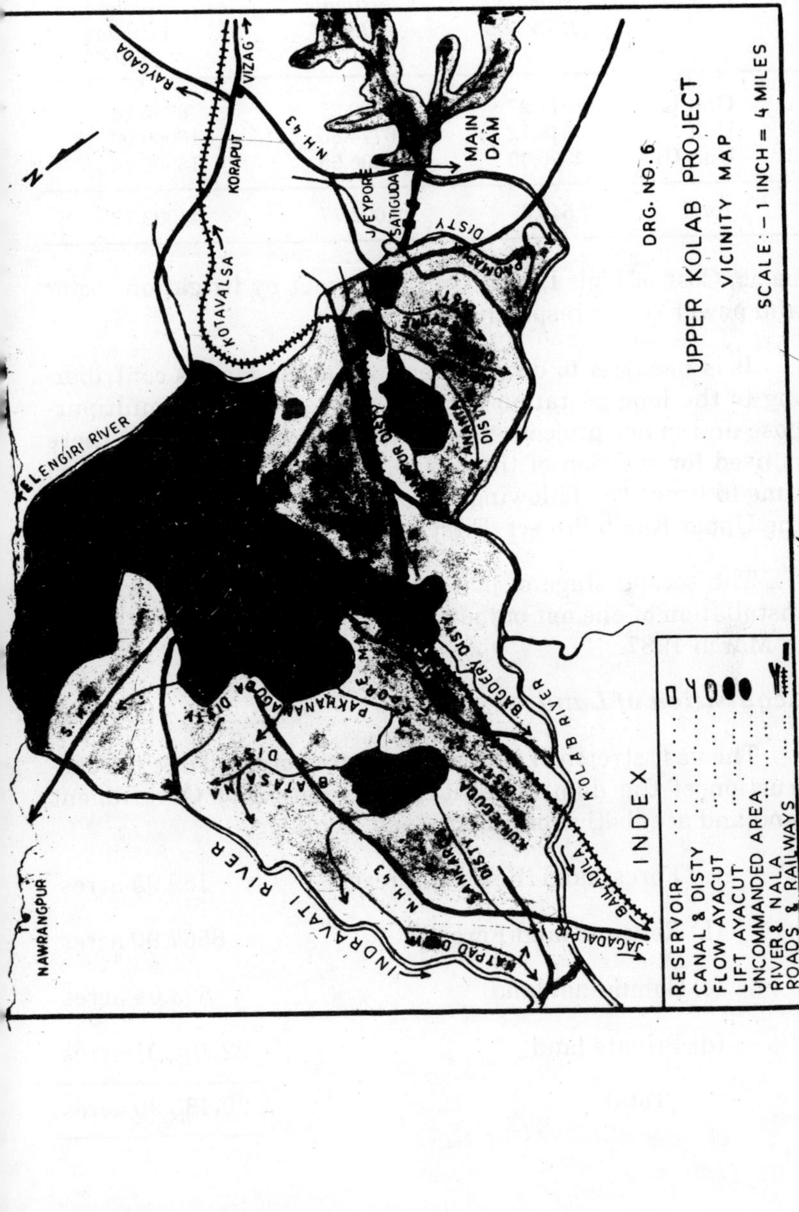
DRG. NO. 6
UPPER KOLAB PROJECT
VICINITY MAP
SCALE :- 1 INCH = 4 MILES
INDEX
RESERVOIR
CANAL & DISTY
FLOW AYACUT
LIFT AYACUT
UNCOMMANDED AREA
RIVER & NALA
ROADS & RAILWAYS
MAIN DAM
KOLAB RIVER
INDRAVATI RIVER
TELENGIRI RIVER
NAWRANGPUR
JAGADALPUR
KORAPUT
VIZAG
RAYGADA
N.H. 43
KOTAVALSA
JEYPORE
SATIGUDA
DISTY
BAGDERY DISTY
N

Map

(Rs. in lakh)

S. No.	*Head of expenditure.*	*Original estimate (1975-76)*	*Ist Revised estimate (1979)*	*Latest revised estimate (1987)*
1.	Unit I	1517.27	5790.27	9762.28
2.	Unit II	1646.12	6774.85	16000.00
3.	Unit III	4380.00	9506.65	18306.44
	Total	7,543.39	22,071.77	44068.72

basis. Cost of Unit II and Unit III is met by irrigation sector and power sector respectively in full.

It is needless to discuss here about the factors contributing to the long gestation period for construction of multipurpose and major projects. Due to cost and time over run there is need for revision of the estimated cost of the project from time to time. The following statement will give a brief idea of the Upper Kolab Project (First stage).

The second stage of power development which relates to installation of one out of four units costs Rs. 1861.73 lakhs as in March 1987.

Acquisition of Land

The vast stretch of water spread area created due to construction of the dam necessitated alienation of Government land and acquisition of private land as under:-

(a) Forest land (Reserve forest)	189.95 acres
(b) Revenue land(Forest)	6567.90 acres
(c) Communal land	533.04 acres
(d) Private land	22,195.01 acres
Total	29,485.90 acres

Rehabilitation

The project displaced 3171 families of whom 1421 belong to S.T., 442 to S.C. and 1308 to General Community. Keeping the rehabilitation policy of the State, then in force, in view, 528 families have seen shifted to project sponsored colonies. Remaining 2643 families have been given cash grant in lieu of land. Besides there was need for rehabilitation of roads and institutional buildings. The subject of study being human resettlement, it is necessary to discuss land acquisition and displacement of human settlement and their rehabilitation in greater detail in the following chapters.

Acquisition of Land

Procedure

The Land Acquisition Act, 1894, is a Central Act under which land for public purpose is acquired. It was amended in 1984. The salient features of the Act are as under:

(*a*) The requiring officer (in the project the Executive Engineer incharge of the Development Division) gives requisition to the Collector for acquisition of land required for the Project. (The Spl. L.A.O. has been declared as the Collector under the Act for this limited purpose).

(*b*) (*i*) The Collector, on receipt of the requisition, sends draft notification under section 4(1) of the Act to the State Revenue Department for publication of the same under intimation to the Administrative Department and the Requisitioning Officer.

(*ii*) If the land is urgently required in public interest, emergency under section 17(4) of the Act is applied. In that event, provisions of the Act as laid down under section 5A, for inviting objections against the acquisition gets omitted.

(*c*) The Government in Revenue Department, on receipt of draft notification under section 4(1) and emergency clause, if any, under section 17(4) cause publication of the said notification in an extraordinary issue of the Orissa Gazette of the L.A. Amendment Act, 1984. The 4(1) notification also gets published in two daily papers circulating in the locality, of which one should be in the regional language and the Collector is also required to get a public notice containing the substance of the said notification served in the village expeditiously. The dates of publication of that notification in the official gazette/ news papers and the date giving public notice shall be deemed to be the date of publication of the notification.

(*d*) After publication of the notification under section 4(1) of the Act, the Collector submits draft declaration papers under section 6(1) of the Act, which consists of a draft declaration, Land Acquisition estimate, land schedule, land plan and grounds of valuation etc., to the concerned R.D.C. and simultaneously, a copy be communicated to the Administrative Department as well as Revenue Department for sanction of L.A. estimate and issue of draft declaration U/S 6(1).

(*e*) On receipt of the D.D (Draft Declaration) papers and L.A. estimate from the R.D.C., the Administrative Department sanctions the L.A. estimate if funds under the appropriate head of account is available. It then sends the D.D. papers and estimate to the Revenue Department for publication U/S 6(1) of the Act.

(*f*) (*i*) Government in Revenue Department, on project of the D.D. papers from the Administrative Department, publish the same in an extraordinary issue of the Orissa Gazette. The declaration is also required to be published in two daily newspapers circulating in the locality, of which, one should be at least in the regional language. Public notice shall also be served by the Collector as in the case of 4(1) notification.

(*ii*) The notifications and declarations are usually published in the Government Press. As delay occurs to a great extent on this account, and the project work suffers, the L.A. Notifications and declarations are also published in Private Press after obtaining concurrence of Revenue Department and Commerce Department.

(iii) The period of validation of notification U/S 4(1) of the L.A.Act is one year from the date of publication. In case the 4(1) notification is not followed up by issue of 6(1) declaration within one year, the 4(1) notification gets invalidated.

(*g*) As soon as the declaration U/S 6(1) is published, and a copy of the public notice so served in the village by the Collector is received by the Government in Revenue Department, an order U/S 7 of the L.A..Act is issued by the Revenue Department authorising the Collector to take over from the tenant and make over possession of the acquired area to the requiring authority.

(*h*) Under Section 11 of the L.A.Act, the Collector makes award of compensation payable to the land owners towards the area acquired.

(*i*) The entire L.A. proceedings shall automatically lapse if no award is made by the Collector U/S 11 within a period of two years from the date of issue of declaration U/S 6(1) by the Revenue Department.

(*j*) If in any case declaration has been published before 24.9.84, award has to be made within a period of two years from 24.9.84. In other words, if a declaration has been published before 24.9.84 but award could not be made by 23.9.84, the entire L.A. proceedings shall also be lapsed.

(*k*) After observing all formalities as provided under sections 12 to 15, the Collector takes over possession of the land under section 16 of the L.A.Act which shall thereupon vest absolutely in the Government free from all encumbrances and makes over the same to the Requiring authority.

Determination of Market Value of Land

(*a*) Preparation of Land Acquisition estimate is not an easy task. For this, the market value of the land proposed to be acquired is to be determined. Normally this is determined after taking into account the sale transactions at the date of publication of notification U/S 4(1) of the L.A. Act. In the event of non-availability of such data, the market value of the land prevailing on or close to the date of submission of the proposal may be taken into consideration.

(*b*) If no such sale transaction data are available, sales of such quality of land in the vicinity within a fairly recent date can be taken into account.

(*c*) If no such statistics as contemplated in (*b*) above are available, the net annual produce of the land can be taken into account for valuation.

(*d*) Value awarded in previous land acquisition proceedings is also relevant evidence on the question of market value.

(*e*) For determination of market value of agricultural land, the annual letting value is also taken into account. The net profit the tenant derives from the land over a certain number of years is taken as the letting value.

(*f*) While determining the market value of the land, value of trees, houses, wells, tanks, crops, etc., that are standing on the land shall be taken into account.

Valuation

(*a*) In addition to the market value of land, additional market value at the rate of 12 per cent of the said market value shall be paid in every case for a period from publication of notification U/S 4(1) to the date of award or the date of taking over possession of the land whichever is earlier.

(*b*) Solatium at the rate of 30 per cent of the original market value of the land shall also the added in every case.

(*c*) Capitalised value of 25 years purchase of the annual revenue is to be provided in the preliminary estimate in all irrigation, power and multipurpose projects which are treated as commercial projects, except for Irrigation Research Institute and Flood Control works.

(*d*) Besides the above, 20 per cent of the estimated compensation amount of land, towards establishment cost, news—paper publication expenses, land charges and contingencies will be realised from projects not bearing expenses of Spl. Land Acquisition Organisation. Where the projects bear the expenses of Spl. Land Acquisition Organisation, 10 per cent of the estimated compensation amount of the land will be realised and added in the estimate.

Interest

In every L.A. Case where possession has been taken after 30.4.82, but payment has not been affected, interest at the rate of 9 per cent shall be paid from the date of taking such possession till the date of payment and interest at the rate of 15 per cent per annum would be paid beyond one year from the date of taking such possession till the date of payment.

Land Acquisition in U.K. Project

Preparation

The F.R.L. (Full Reservoir Level) of the U.K. reservoir is 858.00 metres. It is also the M.W.L.(Maximum Water Level). It was, therefore, decided to acquire land, structures, etc., and remove habitation upto EL 858.00. The work of F.R.L. in the village map after carrying levels in the field, commenced in October 1975 and it was completed in January 1978. Thus, after detailed survey it was assessed that the reservoir would submerge the following:-

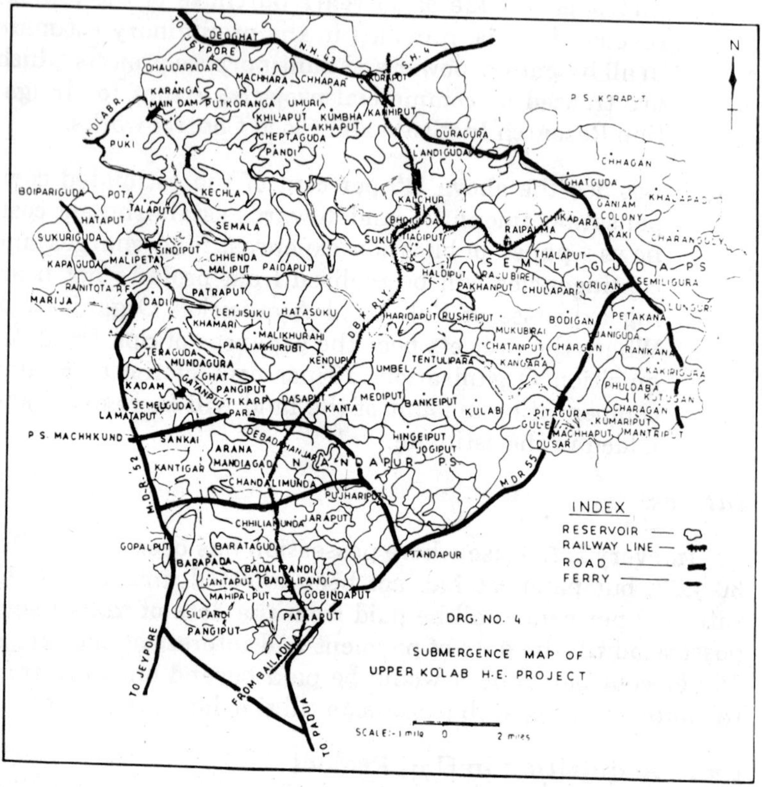
TO JEYPORE
DEOGHAT
N.H. 43
S.H. 4
N
P.S. KORAPUT
DHAUDAPADAR
MACHHARA
CHHAPAR
KORAPUT
KARANGA
MAIN DAM
PUTKORANGA
UMURI
KHILIAPUT
KUMBHA
KANHIPUT
LAKHAPUT
CHEPTAGUDA
PANDI
DURAGURA
LANDIGUDA
KOLAB R.
PUKI
CHHAGAN
GHATGUDA
KECHLA
KALCHUR
GANIAM COLONY
BOIPARIGUDA
POTA
TALAPUT
HATAPUT
SEMALA
KAKI
RAIPALMA
SUKURIGUDA
SINDIPUT
CHHENDA MALIPUT
PAIDAPUT
THALAPUT
KAPAGUDA
MALIPETA
HALDIPUT
RAJUBIRET
S E M I L I G U D A P S
RANITOTA R.F.
PATRAPUT
PAKHANPUT
CHULAPARI
KORIGAN
SEMILIGURA
MARIJA
DADI
LEHJISUKU
HATASUKU
HARIDAPUT
RUSHEIPUT
BODIGAN
PETAKANA
KHAMARI
D.B.K. RLY.
MUKUBIRAI
MALIKHURAHI
CHATANPUT
CHARGAN
RANIKANA
PARAJAKHURUBI
KANGARA
TERAGUDA
KENDUPUT
TENTULIPARA
MUNDAGURA
UMBEL
GHAT
PHULDABA
KADAM
GATANPUT
PANGIPUT
TIKARP
DASAPUT
MEDIPUT
SEMELGUDA
PARA
KANTA
BANKEIPUT
KULAB
PITAGURA
CHARAGAN
LAMATAPUT
KUMARIPUT
GULE
MACHHAPUT
MANTRIPUT
P.S. MACHHKUND
SANKAI
DEBADA
HINGEIPUT
JOGIPUT
DUSAR
ARANA
N A N D A P U R P.S.
M.D.R. 55
KANTIGAR
MANDIAGADA
CHANDALIMUNDA
M.D.R. 52
PUJHARIPUT
INDEX
RESERVOIR
RAILWAY LINE
ROAD
FERRY
CHHILIAMUNDA
JARAPUT
GOPALPUT
BARATAGUDA
MANDAPUR
BARAPADA
BADALIPANDI
JANTAPUT
MAHIPALPUT
GOBINDAPUT
SILPANDI
PATRAPUT
DRG. NO. 4
PANGIPUT
SUBMERGENCE MAP OF
UPPER KOLAB H.E. PROJECT
TO JEYPORE
FROM BAILADILA
TO PADUA
SCALE:- 1 mile 0 2 miles

Map

(*a*) Private land 23,856.05 acres

(*b*) Reserve Forest land 189.95 acres

(*c*) Revenue forest land 6,567.90 acres

(*d*) Communal land 533.04 acres

(*e*) Roads

(i) National Highway 8km

(ii) Major District Road At different places.

(iii) Water Works of Public Health Department.

(iv) High Voltage transmission line of OSEB.

(v) Properties of Central Cattle Breeding Farm and Orissa Small Industries Corporation, Sunabeda.

Drawing No.4 gives an idea about the extent of submergence by Upper Kolab Reservoir.

The Executive Engineer in charge of the work was declared as the requiring officer for the lands inside the reservoir.

Land Acquisition Organisation

Keeping the magnitude of land acquisition work of the project and on the line of requirement of multipurpose project in view, a Special Land Acquisition Officer has been posted for the project. He is also incharge of the resettlement and rehabilitation work. As such he is designated as the Spl.Land Acquisition and Resettlement Officer. Although the project is located about 5 Kms away from the N.H.42, half way between Koraput and Jeypore, the Spl.L.A and R.O.holds his office at Koraput. The details of the peak sanctioned strength of staff of the Spl.L.A and R.O. are indicated in Table 5.1.

The Spl.L.A. and R.O. is a Class I Officer of the Orissa Administrative Service and the Zone Officers and Special Officer are Class II O.A.S. Officers.

Table 5.1
Details of Peak Sanctioned Strength of Staff of Land Acquisition and Resettlement Officer (Para 5.4.02).

S. No.	*Designation*	*No. of posts.*
	Gazetted.	
1	L.A & R.O (O.A.S.I)	1
2	Zone Officer (O.A.S Class II)	15
3	Special Officer (O.A.S.Class II)	1
		17
	Non-Gazetted	
1.	Officer On Speical Duty Rehabilitation/Head Asst.	1
2.	Revenue Inspector	28
3.	Amin	140
4.	Process Server	30
5.	Senior Clerk	3
6.	Junior Clerk	9
7.	Draughtsmen	6
8.	Jr.Accountant	1
9.	Copyist	1
10.	Nazir	1
11.	Senior Stenographer	1
12.	Driver	4
13.	Agricultural Overseer	2
14.	Junior Clerk-cum-Nazir	1
		228
	Class IV	
1.	Chainman	143
2.	Peon	24
3.	Daftry	1
4.	Night Watchman	14
		182
	Total	427

Acquisition of Forest Land

Under Forest (Conservation) Act, 1980, clearance from Government of India, in the Ministry of Forest and Environment is necessary for diversion of forest land for non-forest uses. But since Upper Kolab Project was administratively approved in 1976, the aforesaid Act did not have any bearing on the project clearance.

Reserve Forest Land

189.95 acres of reserve forest land was submerged in the reservoir in 27 villages. For raising compensatory afforestation as per rules, Rs.2,15,697.20 P was paid to the State Forest Department.

Revenue Forest

6567.90 acres of miscellaneous forest land recorded in the name of Revenue Department were submerged in the three tehsils of Pottangi, Koraput and Machkund. As per rules these lands were formally alienated in favour of Irrigation and Power Department for eventual submergence in the reservoir.

Acquisition of Private Land

In total 23856.05 acres of private land in 142 number of villages was acquired, of which 631.64 acres are communal land and 303.83 acres are homestead land. The rest are agricultural land. The procedure as laid down in the Land Acquisition Act, 1894 (amendment Act 1984) was adopted in this regard.

Valuation of Land

Market value of land was calculated after taking the actual sale deed in the locality within three years prior to 4(1) notification and additional market value at the rate of 12 per cent per year for three years into consideration. Further 30 per cent extra towards solatium charges were added. As statistics reveal, the highest rate paid is Rs. 8000 per acre for paddy I class land and the lowest rate paid Rs. 340 per acre for dry land.

Valuation of Trees and Their Disposal

As has been stated previously the extent of reserve forest land was 189.95 acres only. Hence the reservoir area was very sparsely vegetated with a few clusters of trees at and near villages on private land. Value of the trees was paid to the owners. Trees coming under good species were valued basing either on the cost of fruits or on cost of wood. In case the cost of fruit was more than the cost of wood, the cost of fruit was considered and *vice versa*. In case of unspecified trees, the cost was calculated taking them as firewood. Table 5.2 indicates the rates paid for different types of trees.

For the healthy life of the reservoir and the dam and for efficient pisciculture it is necessary that the bed of the reservoir is cleared off all the tree growth. The responsibility of felling the trees and disposing them was entrusted to the Orissa Forest Corporation.

Valuation of Structures

Besides land, house, wells, tanks and other structures were also submerged for which compensation was paid. The norm followed in this regard has been given in Annexure I. It is good that calculating compensation for structures, no de-

Table 5.2

Rates of Different Varieties of Trees for Calculation of Compensation.

S.No.	*Name of trees*	*Unit*	*Rate*
1	*2*	*3*	*4*
1.	Mango	Each	Rs. 90.00
2.	Tamarind	Each	Rs.90.00
3.	Jack-Fruit	Each	Rs.60.00
4.	Karanja	Each	Rs.30.00
5.	Lemon	Each	Rs.30.00
6.	Orange	Each	Rs.30.00
7.	Bamboo	Each	Rs. 2.00
8.	Jafra	Each	Rs.30.00
9.	Simuli	Each	Rs.30.00
10.	Guava	Each	Rs.30.00
11.	Jamun	Each	Rs.30.00
12.	Salap	Each	Rs.120.00
13.	Dumuri	Each	Rs.30.00
14.	Miscellenous i.e., firewood.	Each	Rs.6.00

duction towards depreciation on buildings, wells, etc., has been made.

The details of private land, trees, structures acquired for the project in phases including after resurvey and compensation money paid are furnished in Annexure-II. It would be seen that in total Rs. 6,30,23,389 was paid towards compensation money to the families who were affected by the Upper Kolab Reservoir.

Rehabilitation

The rehabilitation policy adopted for Rengali Multipurpose project from time to time was made applicable to Upper Kolab Project too. But the amendments made in and after 1987, were not adopted because of restrictions imposed on U.K Project and similar many other projects.

Enumeration of Families

The general practice followed in reservoir projects is to identify all the families who need to be evacuated from the area coming below the Full Reservoir Level. In case of U.K. Project initially it was estimated that about 6400 families were to be affected by the reservoir in 147 villages.Family in relation to a displaced person as per the rehabilitation policy was such person and his/her spouse, minor sons, unmarried daughters, minor brothers and sisters, father, mother and other members residing with him and dependent on him for their livelihood.The reference date for enumeration of family was 11th June 1975 *i.e.,* the date on which the foundation stone for the U.K. Project was laid. But in reality 2972 number of families belonging to 49 number of villages and 2 hamlets who were identified to have been affected were displaced in phases by June 1987

Again, the assessment made as above was not correct. F.R.L. 858M was incorrectly demarcated in the field. F.R.L. of 858 M to have been achieved during the monsoon of 1987 was actually 856 M. Even at 856M level, additional land area was found to have been submerged thereby affecting more number of familieo. Hence immediately after this mistake was detected, fresh survey work was conducated to find out the correct position. Fresh survey in June 1988 revealed that additional 473 number of families in nine number of new villages would be affected, of whom 95 number of families only needed rehabilitation. Another recheck survey in June 1989 also revealed that 104 additional families in two villages from out of above nine villages required to be shifted.

Thus, in total (2972+95+104) 3171 number of families were identified for evacuation. The details of displacement of families have been discussed in succeeding paragraph.

Land for the Oustees

"Land for Land" is the guiding principle for resettling and rehabilitating the displaced persons. It was the primary responsibility of the project authorities to locate adequate extent of land both for homestead purposes and agricultural purposes and to keep them in order by the time physical displacement took place. About 20,000 hectares of land would have been required for this purpose. But it was assumed that all the displaced persons would not move to the land so located by the project authorities. Hence requirement was assessed on a lower side.

Identification of Land

Hardly in any project in Orissa, private land has been acquired for resettling the oustees. In case of Upper Kolab Project Government land as follows was made available to the project authorities for resettlement purposes.

(*a*)	Batasana and Nuagaon Circles in 23 villages of Kotpad Tehsil	..9432.00 acres
(*b*)	Boriguma Circle in 25 villages of Boriguma Tehsil	...3248.00 acres
		12,680.00 acres

Table 5.3
Details of Land made Available for Resettlement of Oustees

Area shown in acres

S.No.	*Village*	*Classification of land* *Forest*	*Patra Jungle*	*Scrub*	*Others*	*Total*	*Area reclaimed*
1	2	3	4	5	6	7	8
1.	Batasana	3285.28	-	-	907.14	4192.42	3694.00
2.	Amta	-	738.16	-	85.90	824.06	824.06
3.	Tergia	-	97.82	-	163.80	261.62	176.78
4.	Bukubandha	-	190.65	-	21.01	211.66	140.00
5.	Chitragaon	-	146.75	-	-	146.75	120.00
6.	Dhanpur	-	106.51	-	-	106.51	106.51
7.	Miriguda	-	154.60	-	3.26	157.86	114.00
8.	Bangiguda	-	53.62	-	-	53.62	53.00
9.	Sargiguda	-	47.60	-	-	47.60	47.60
10.	Chatarla	71.68	-	-	-	71.68	37.55
11.	Nuagaon	-	138.07	-	935.22	1073.29	645.00
12.	Ekamba	-	79.19	-	33.10	112.29	112.29

(contd.)

Table 5.3 (Contd.)
Details of Land made Available for Resettlement of Oustees

1	2	3	4	5	6	7	8
13.	Ghumar	279.79	-	-	13.96	293.75	293.75
14.	Nehrupali	-	527.13	-	-	527.13	255.00
15.	Hardoli	-	278.14	-	100.00	378.14	150.00
16.	Phupugaon	-	140.29	35.20	-	175.49	90.00
17.	Khumahandi	-	145.97	-	-	145.97	40.00
18.	Bastamba	-	48.84	-	-	48.84	48.00
19.	Bandhugaon	92.50	-	-	15.25	107.75	65.00
20.	Birahandi	-	11.31	121.09	-	132.40	44.00
21	Ambavalli	-	62.54	-	25.30	87.84	55.00
22.	Bansoli	84.94	-	-	36.44	121.38	38.00
23.	Masigaon	131.75	-	-	23.20	154.95	154.95
Total		3945.94	2967.19	156.29	2363.58	9433.00	7304.49

(In Boriguma circle also 3248.00 acres of land in 25 villages were handed over to the project authorities. But neither the land there was reclaimed nor any resettlement activities were taken up).

Details of the land made available for resettlement are furnished in Table 5.3.

Reclamation of Land

The Rehabilitation policy clearly speaks that the land for both homestead and agricultural purposes should be reclaimed before allotment to the oustees. Accordingly the land so made available is required to be reclaimed. In the first phase, land in Batasana and Nuagaon Circle was reclaimed. Of this 7,609.42 acres was forest land. At the relevant point of time there was no restriction to get forest land for such purposes. The land in Boriguma circle was not reclaimed as the response from the oustees to settle there was negative.

At the initial stage, the Reclamation and Rehabilitation Organisation of Dandakarnya Development Authority undertook the reclamation work. During the period from May 1979 to June 1980, they could complete re-clamation work in 1318 acres only. The said organisation shifted its activities elsewhere; for which the reclamation work was done manually through various agencies. Table 5.3 also indicates the extent of land reclaimed in different villages. Out of the total extent of 9,433 acres available, 6,455.68 acres needed reclamation.

Displacement

As per the construction programme chalked out for the project, the Upper Kolab reservoir was to be impounded in three phases upto elevation 835M, 850M and F.R.L.858M. Accordingly the work of displacement of persons was taken up.

(a) First Phase of Displacement

As per programme, the water level in the reservoir was to rise upto EL 835M during the monsoon of 1984. Five villages involving 694 families were to be affected. All arrangements were made to evacuate these families to the colonies set up by the project. But only 24 families went there and the rest made their own arrangement after receiving cash grant in lieu of land. Some shifted to higher altitudes and some to

Kolabnagar to work in the project activities.

(b) Second Phase of Displacement

Water level in the reservoir was to rise upto 850M affecting thereby 21 villages with 1,410 families during the monsoon of 1985. Of them only 58 families opted to go to the colonies and the rest received cash. The work was completed by July-August 1985. Water level in the reservoir then was 845.00M only.

(c) Third Phase of Displacement

The third phase of work was originally programmed to be completed during the monsoons of 1986 before the water level reached the F.R.L. of 858M. But due to non-completion of several relocation works by State P.W.D. it was decided to impound water in the reservoir upto EL 850M only. However, prior to it 78 number of families in five villages located above EL 850M were evacuated in June 1985 after payment of full cash grant. Later in June 1987, 710 number of families belonging to 18 villages including 2 villages from aforesaid five villages were shifted. Of them 301 families shifted to the colonies and the rest received cash grant to make their own rehabilitation arrangements.

(d) Forth Phase of Displacement

During the monsoon of 1987, the reservoir level rose upto 856M. In all fairness all the land coming below the F.R.L. of 858M should have been acquired and the families coming within that area after due enumeration should have been displaced. But as ill luck would have it, due to incorrect survey, the F.R.L. was not demarcated in the field properly and for which several families coming under submergence area were not counted previously. However, fresh survey work was undertaken to find out extra land to be acquired and additional families affected to be evacuated. The resurvey in June 1988 revealed that in nine number of more villages, 970.40 acres of extra land would be acquiread and out of 473 families residing in all the nine villages, 95 families needed rehabilitation elsewhere. Subsequently in June 1989, two villages from out of above nine were also notice to be affected involving additional 104 number of families. Accordingly although it was decided to complete the work of displacement in three phases by June 1986, the work still continues as the last batch of 104 families were shifted in 1991.

The above four Phases of displacement relate to the reservoir area only. Besides two hamlets of Jeypore village came under the Satiguda pond created for storing the tail race discharge water from the power house, for irrigation purposes. Eighty number of families from these two hamlets were displaced. Thus, in total 3171 number of families were displaced; the details of which are given in Annexure-III. It would be seen that out of total 3171 number of families, 528 families shifted (including 104 families still to shift) to colonies set up by Government and remaining 2643 families moved elsewhere on their own. A sum of Rs. 3,88,85,103 was disbursed as rehabilitation grant in cash in lieu of land.

Disturbance in other Sectors

The reservoir not only caused displacement of human settlement, but also caused disturbance to the structures, of several departments and communication system. They are discussed as follows:-

(a) National Highway

(i) 8 Km length of National Highway No.43 between Koraput and Sunabeda was submerged by the reservoir. Hence the road for this portion was to be realigned at a cost of Rs. 585.89 lakhs, out of which the Upper Kolab Project shared Rs.515.89 lakhs.

(ii) For the Jeypore Main Canal, the N.H. 43 had to be pierced through at two different places near Jeypore. Hence the project had to bear Rs.44.78 lakhs and Rs. 15.82 lakhs towards restoration in shape of construction of bridges.

(b) Major District Road No. 52

The reservoir submerged the M.D.R. 52 at a number of

places between KM 182/0 and Km 186/0. In order to avoid the reservoir, the road was realigned at a cost of Rs. 73.55 lakhs. The project paid this amount to State Works Department (R and B).

Another stretch of M.D.R. 52 between Km 169/6 and Km 170/4 near village Tusuba was also affected by the reservoir. For its realignment the project provided Rs. 34.56 lakhs.

(c) Major District Road No. 55

A portion of the road between Similiguda-Nandapur (Km 8/2 to Km 9/2) was affected by the reservoir. River Kolab was flowing in this stretch of road. Consequently the existing bridge and the road were to be substituted at an estimated cost of Rs. 91.64 lakhs. The work is being done by State Works Department (R and B). During the interim period, ferry service was run.

(d) Public Health Department

The intake works of Jawahar Water Works, Sunabeda and Koraput Water Works were submerged by the reservoir, necessitating their relocation. Relocation and ancillary works were completed by the State P.H.D at a cost of Rs. 195.26 lakhs and Rs. 44.00 lakhs; of which the U.K Project shared Rs. 132.25 lakhs and Rs. 20.67 lakhs respectively.

(e) Orissa State Electricity Board

The high voltage transmission lines of O.S.E.B were submerged at different places. Hence the U.K. Project paid Rs.8.66 lakhs to OSEB towards relocation of those lines.

(f) Forest Department

189.95 acres of Reserve Forest Land was submerged in the reservoir, for which the U.K. Project paid Rs. 2.16 lakhs towards compensatory afforestation. For setting up rehabilitation colonies, 5,200 acres of forest land were also reclaimed. Hence a further sum of Rs.4.62 lakhs was paid to Forest Department towards compensatory afforestation. For Revenue forest land no payment had been made as at the relevant time there was no provision for making any payment to Forest Department towards compensatory afforestation.

(g) Central Cattle Breeding Farm

The reservoir submerged the existing internal road of the farm. Realignment work was done at the project cost.

(h) Orissa Small Industries Corporation

The Carpentary Workshop and staff quarters of the Corporation at Sunabeda were submerged, for which the project gave compensation of Rs. 12.375 lakhs.

(i) School and Hospital

One Kanyashram (High School) which came under submergence was shifted from the reservoir area by constructing a new school. Hospital and staff quarters were also constructed near the school in village Deoghati.

The details of relocation work and payment of compensation money are given in Table 5.4.

21 L.P.Schools, 4 U.P. Schools, one P.S.M.E. School, one temple and one Panchayat Ghar, one Grain Gola, One Livestock Centre, One V.L.W. Centre were also affected by the reservoir. But excepting the Panchayat Ghar, livestock Centre and V.L.W. Centre (previously located in Malipondi village shifted to Lakhanput and P.S.M.E. School at Malipondi to Pujariput in Koraput N.A.C.) the rest have not been relocated so far. The details of these institutions are given in Table 5.5.

Rehabilitation Works

The three major items of rehabilitation measures provided under the State Rehabilitation Policy are land, cash payment and employment. Works done in this regard are discussed in the following:-

(*a*) *Land*- "Land for land" is the guiding principle of re-

Table-5.4

Details of Payment Made to Different Departments/ Organisations Towards Compensation for Their Property Submerged in Upper Kolab Reservoir

(in lakh rupees)

Sl. No.	*Item of work*	*Estimated cost / Project share*	*Amount paid*
1.	Semiliguda-Handiput Road (Bridge with approach 5 Km) MDR-55.	91.64 / 91.64	91.64
2.	Vizag-Jeypore Road MDR-52 (Raising 200 M with bridge).	26.304 / 26.304	30.00
3.	Vizag-Jeypore Road MDR-52 (Realignment 4 Km)	73.550 / 73.550	73.550
4.	N.H.43 (I&P share)	585.89 / 515.80	515.89
5.	Jawahar Water Works Sunabeda.	152.40 / 132.25	132.25
6.	Koraput Water Works	44.00 / 20.67	20.67
7.	O.S.E.B. (33 K.V. & 11 K.V. line) (Diversion of 33 KV line in between Gopalpur to Sunabeda.	8.139	6.064
8.	Construction of 33 KV line near OSAP 3rd Bn. Company at Jhatiguda dyke.	1.648 / 1.648	1.648
9.	Construction of new 11 KV line near OSAP.	0.954 / 0.954	0.954
10.	Orissa Small Industries Corporation, Sunabeda	12.375 / 12.375	12.375
11.	Posts & Telegraph	0.356 / 0.356	0.356
	Total	997.256 / 883.686	885.397

Table 5.5
Details of Institutions

S.No.	Village	L.P. school	U.P. school	PSME school	Club house	VLW Centre	Pan-chayat Centre	Live stock Centre	VLW Centre	Grain gola
1.	Gopalput	2	1	-	1	-	-	-	-	-
2.	Koranga	3								
3.	Thalaput	1								
4.	Kechela	1								
5.	Balda	1								
6.	Dhadi	-	-	-	-	-	1	-	-	-
7.	Dumuriguda	1								
8.	Haldiput	1								
9.	Malipeta	1								
10.	Rajubedei	1								
11.	Khapraput	-	1							
12.	Khilaput	1	-	-	1					
13.	Tiadiput	-	1	-	-	-	One temple			
14.	Landiguda	1								
15.	Kalchuru	1								
16.	Bhoiguda	1								
17.	Sirisi	1								
18.	Malipondi	-	-	1	-	1	1	1	1	1
19.	Parojapondi	1								
20.	Paidaput	1								
21.	Puki	1								
22.	Kulaba	-	1							
	Total	21	4	1	2	1	2	1	1	1

settlement of oustees of an irrigation project. It has been discussed 6 that 12,680 acres of land were set apart for allotment to the oustees for homestead purposes as well as for agricultural purposes. Homestead land at the rate of 0.5 acre and unirrigated land at the rate of 6 acres or irrigated land at the rate of 3 acres per family (as prevalent then) was to be given. Preliminary assessment revealed that about 25 per cent of displaced people would be interested to move into the land so set apart from them. Hence the land available would have been adequate to meet the requirement. But there was very poor response from the oustees. We shall discuss about this later. Hence reclamation work was confined to 9433 acres of land available in Batasana and Nuagaon Circle only to accommodate 2072 number of families. Accordingly seven number of colonies were set up there. Details of these colonies are given in Table 5.6.

(a) House Construction

In keeping with the then prevailing provisions of Rehabilitation policy, forest materials for construction of houses by the oustees were provided by the Forest Department at single royalty tenants' rate applicable for "Khesra" forests. They were supplied at the following scale per family:-

(i)	Poles 75 Cms to under 90 Cms -	10 Nos.
(ii)	Poles under 60 Cms -	5 Nos.
(iii)	Poles 30 Cms to 45 Cms -	4 Nos.
(iv)	Bamboo or Poles under 30 Cms -	50 Nos.
(v)	Brush wood -	2 cart loads

(vi) One tree of 120 Cms girth per family at the prescribed rate.

(b) Common Facilities

Table 5.6
Various Rehabiliation Works done Seven Colonies.

S.No.	Items	Colonies					Total	
		No. 1 & 2	No.3	No.4	No.5	No.6	No.7	
1.	No. of families							
(a)	Programmed to be resettled.	156	118	687	248	228	585	2,072
(b)	Actually resettled	nil	nil	241	56	31	96	424
2.	Area allotted for different purposes (in acres)							
(a)	Irrigable land	340.00	232.00	1147.00	475.00	275.00	479.00	2,948.00
(b)	Non-irrigable land.	222.00	7.00	682.00	257.00	327.00	1501.00	2,996.00
(c)	Homestand land	78.00	59.00	344.00	124.00	114.00	293.00	1,012.00
(d)	Communal land	124.00	155.00	647.00	291.00	383.00	869.00	2,469.00
	Total	764.00	453.00	2820.00	1147.00	1099.00	3,412.00	9425.00
3.	Infrastructure facilities.							
(a)	Roads (in Km)							
(i)	Programmed.	9.5	5.8	10.55	5.00	7.30	14.60	52.75
(ii)	Physically done	9.5	5.8	24.25	8.10	6.30	20.10	74.05
(b)	Schools							
(i)	Programmed	1	1	1	1	1	1	6
(ii)	Physically done	1	1	1	1	1	1	6

(contd.)

Table 5.6 (contd.)

(c) Wells						
(i) Programmed	3	2	6	3	3	7	24
(ii) Physically done.	1	1	2	3	2	2	11
(d) Tube Wells.							
(i) Programmed	0	0	0	0	0	0	0
(ii) Physically done	4	3	3	1	3	1	8
(e) Tanks							
(i) Programmed	1	1	3	1	1	2	9
(ii) Physically done	1	1	3	1	1	0	7

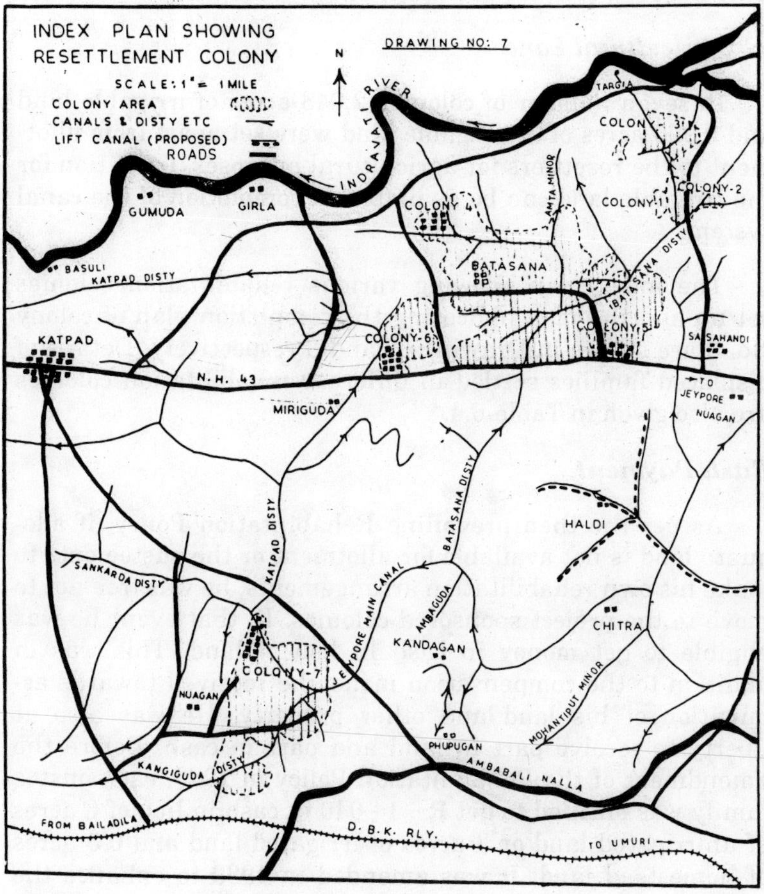
INDEX PLAN SHOWING
RESETTLEMENT COLONY
DRAWING NO: 7
N
SCALE: 1" = 1 MILE
COLONY AREA
CANALS & DISTY ETC
LIFT CANAL (PROPOSED)
ROADS
INDRAVATI RIVER
TARGIA
COLONY-3
COLONY-2
COLONY-1
COLONY 4
AMTA MINOR
GUMUDA
GIRLA
BASULI
KATPAD DISTY
BATASANA
BATASANA DISTY
COLONY-5
COLONY-6
SARSAHANDI
KATPAD
N.H. 43
TO JEYPORE
NUAGAN
MIRIGUDA
HALDI
KATPAD DISTY
BATASANA DISTY
SANKARDA DISTY
JEYPORE MAIN CANAL
AMBAGUDA
CHITRA
KANDAGAN
COLONY-7
NUAGAN
MOHANTIPUT MINOR
PHUPUGAN
AMBABALI NALLA
KANGIGUDA DISTY
FROM BAILADILA
D.B.K. RLY.
TO UMURI

Map

Common facilities *e.g.*, village roads, schools wells, tubewells and tanks were also provided at project cost. Besides two High Schools, one at Sashahandi and the other at Nuagaon were set up. The details of the rehabilitation works done in seven colonies are indicated in Table 5.6.

(c) Agricultural Land

In seven number of colonies 2,948 acres of irrigable land and 2,996 acres of unirrigable land were set apart from allotment to the resettlers for agricultural purposes. Irrigation for the irrigable land can be available on completion of the canal system.

The index plan showing various rehabilitation colonies set up and the map indicating the completion plan of colony No. 3 are shown in Drawings 7 and 8 respectively. Details of displaced families settled in different rehabilitation colonies are also given in Table-6.4.

Cash Payment

As per the then prevailing Rehabilitation Policy, if adequate land is not available for allotment or the oustee opts to make his own rehabilitation arrangements, he was free not to move to the project sponsored colonies. In that event he was eligible to get money in cash in lieu of land. This was in addition to the compensation money he received towards acquisition of his land and other property. He was also at liberty to receive part in land and part in cash. Before the amendment of the Rehabilitation Policy in 1989, each oustee family was entitled to get Rs. 14,040 in cash in lieu of 6 acres of unirrigated land or 3 acres of irrigated land and 0.5 acres of homestead land. It was amended in 1989 to enhance the monetary ceiling to Rs. 20.075 and to reduce the land ceiling to 5 acres of unirrigated land or 2.5 acres of irrigated land. No change was made in the ceiling of homestead land. Further amendment was made in Irrigation Department Resolution No. 31888 dt. 21.8.90 to grant cash at the rate of 8000 per acre of unirrigated land and Rs.16,000 per acre of irrigated land subject to his entitlement. Both the aforesaid

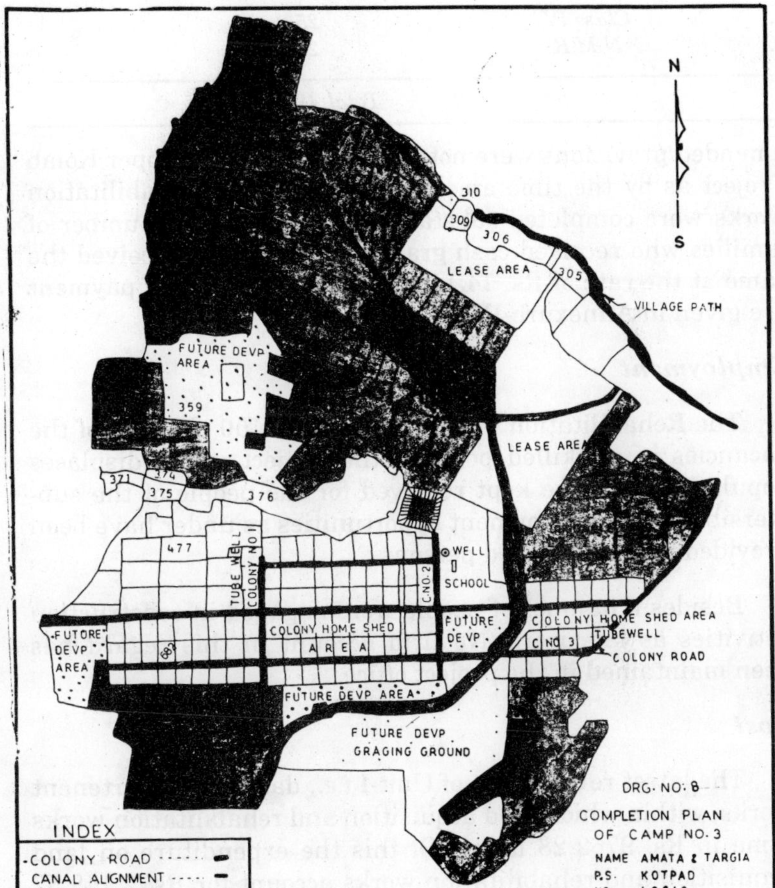
N
S
310
309
306
305
LEASE AREA
VILLAGE PATH
FUTURE DEVP AREA
359
LEASE AREA
371
374
375
376
477
TUBE WELL
COLONY NO 1
WELL
SCHOOL
C.NO.2
FUTURE DEVP
COLONY HOME SHED AREA
TUBEWELL
C NO 3
COLONY ROAD
FUTURE DEVP AREA
COLONY HOME SHED AREA
FUTURE DEVP AREA
FUTURE DEVP GRAGING GROUND
DRG. NO: 8
COMPLETION PLAN OF CAMP NO. 3
NAME AMATA & TARGIA
P.S. KOTPAD
NO. 42843
DIST. KORAPUT
INDEX
COLONY ROAD
CANAL ALIGNMENT
IRRIGABLE AREA
NON IRRIGABLE AREA

Map

S.No.	*Grade of Employment*	*No. of persons employed.*
1.	*Class IV*	*25*
2.	*N.M.R*	*23*
		Total 48

amended provisions were not made applicable to Upper Kolab Project as by the time amendment was made, rehabilitation works were completed substantially. Hence 2,643 number of families who received cash grant in lieu of land, received the same at the rate of Rs. 14,140 per family. Details of payment are given in Annexure-III.

Employment

The Rehabilitation Policy outlines that 50 per cent of the vacancies in unskilled posts in the Project which displaces population, shall be kept reserved for the people of the submersible area. Employment opportunities as under have been provided to the displaced persons.

Besides a number of people were engaged in construction activities as labourers. No firm account in this regard has been maintained in the project office.

Cost

The latest revised cost of Unit-I *i.e.,* dam and appurtenent works, within which land acquisition and rehabilitation works come in Rs. 9762.28 lakhs. Of this the expenditure on land acquisition and rehabilitation works account for Rs. 2,148.57 lakhs. The year-wise breakup of the said expenditure is given in Table 6.18.

Organisation

The Spl. Land Acquisition Officer detailed for ensuring acquisition of land required for the Project is also the Rehabilitation Officer. So far as rehabilitation work is concerned he is responsible to allocate land and other benefits to the displaced persons as provided in the Rehabilitation Policy and to ensure that the affected persons are evacuated as per schedule after payment of due compensation money.

In this work he is assisted by the Executive Engineer of Upper Kolab Development Division and his staff. Reclamation of the land for homestead purposes as well as agricultural purposes, construction of approach roads and internal roads in the settlement colonies, buildings, wells and tanks and miscellaneous civic amenities are the responsibility of the Development Division. This Division was manned in its full strength by One Executive Engineer, 4 Asstt. Engineers, 17 Junior Engineers and several workcharged and seasonal workcharged staff *viz.,* Amin(6), Surveyor (2), Khalasi (13), Watchman (5), Survey Inspector (1), Jeep Driver (5), Truck Driver (3) and Helper (3).

Rehabilitation Advisory Committee

State Government in Irrigation and Power Department Resolution No. 4495-Irr-III(R)5/82 dt. the 9th February 1978, constituted a Rehabilitation Advisory Committee to coordinate the implementation of various schemes under the rehabilitation programmes. The functions of the Committee are -

(*a*) Examine the rehabilitation schemes prepared by the project officials and furnish recommendations to Government.

(*b*) inspect or examine the areas earmarked for rehabilitation of displaced persons and recommend to Government.

(*c*) inspect or examine the details of area to be submerged and decide which of the villages to be fully rehabilitated and which others partly.

(*d*) examine the details of landless and homeless families and approve the same.

(*e*) decide and recommended to Government regarding nature of common facilities to be provided to the displaced persons in different locations.

(*f*) recommend to Government as to what type of irrigation facilities to be provided for the rehabilitation areas.

(*g*) approve the rehabilitation programme, coordinate the implementation of various schemes under the rehabilitation programme and examine from time to time the progress achieved.

(*h*) shall recommend to Government how to remove the bottlenecks in the implementation of rehabilitation programmes.

The Committee is headed by the Revenue Divisional Commissioner, Southern Division. The other official members are Collector, Koraput, Chief Engineer of U.K. Project, Conservator of Forests, Koraput; S.E., Upper Kolab Dam Circle; Soil Conservation Officer, Koraput; Spl.L.A. and R.O. (Member Secretary). The non-official Members are M.Ps (Lok Sabha) Koraput and Nowrangpur; M.L.As Jeypore, Koraput, Pottangi and Kotpad; Chairman, Panchayat Samities of Koraput, Laxmipur, Nandpur, Similiguda, Pottangi, Narayanpatna, Badhugaon and Dasamanthpur.

The permanent invitees are S.E., Upper Kolab Circle, D.F.Os, Jeypore, Nowrangpur and Rayagada, Sub-Collectors, Jeypore and Koraput.

The Committee is required to meet atleast once in three months. In absence of the R.D.C., the Chief Engineer of U.K. Project is to Chair the meetings.

The Committee met twelve times upto end of 1989 and made several recommendations.

6

Lessons from Experience in Upper Kolab Project

6

Lessons from Experience in Upper Kolab Project

Section - A : GENERAL VIEW

Introducton

Through centuries, the prosperity of State has been depending on the success or failure of paddy which is the principal crop of the State. For want of rains during about eight months in the year, rabi cultivation is also not possible without irrigation. In case of Koraput district, agriculture mostly means growing one crop in a year. Water again is the chief limiting factor there in this regard. On completion of the irrigation system, the Upper Kolab Project will bring water to about 88,740 hectares of thirsty land annually and will usher an era of green revolution. Besides it will lead to the following benefits:

(a) Increase in Double Cropped Area

The soil and climate being favourable, besides kharif crop, rabi crops like paddy, oil seeds, fibre crops will be grown as soon as irrigation is available.

(b) Greater Security of Paddy Crop

Due to vagaries of monsoon, one is not presently sure of a full crop of paddy which is the staple food crop, without irrigation. Hence canal irrigation will ensure a full crop in that area.

(c) Increased Production

Year-round irrigation will increase agricultural yield in the Command Area.

(d) Better Farming

Lack of irrigation has been the main difficulty in the adoption of improved practices in agriculture. But with an assured perennial source of irrigation, agricultural improvements will sure to follow. Application of manure at increased rate, timely sowing of seeds, etc., will be an important change in agricultural practice. Transplanation of paddy brings more yield than broadcast method. Presently nursery sowing which is not widely practised because of lack of irrigation facilities, will be also commonly practised.

(e) Security of Income

Irrigation will ensure assured water suppy and consequently there will be security of income. With the help of irrigation, the oustees in settlement colonies will be able to increase the yield by about three times. This will definitely more than compensate losses of their old habitation.

(f) Benefits of Power

Beneficial use of power needs no highlighting. The project will increase the power potential of Orissa by 240 M.W. especially at a time when there is great demand for it.

Due to situational necessities, most multipurpose projects in Orissa are located in backward areas. Upper Kolab Project is no exception. The inhabitants in this area are very poor and backward economically. But as soon as the twin source of progress *i.e.,* irrigation and power are available, the development of the area will become very quick. As a matter of fact, with the increased activities in the project, population of Koraput, Jeypore has increased. Development in the industrial sector has recently started, NALCO at Damanjodi and SEWA Paper Mills at Jeypore have come up. These industries have already opened and the incoming industries will open employment opportunities. Consequently the per capita income of the people will go up.

The Upper Kolab Project itself during its construction has provided employment to a large number of skilled and unskilled workers. On the civil side, as many as 109 gazetted technical, 183 specially declared gazetted technical, 3 non-technical gazetted, 7 non-technical specially declared gazetted, 347 other non-technical non-gazetted staff declared gazetted, 347 other non-technical non-gazetted staff and 388 Class IV staff were employed. Besides there were 93 staff of different categories in the Office of the Financial Adviser and Chief Accounts Officer, 427 staff in the Office of the Special Land Acquisition and Rehabilitation Officer, 68 staff in the hospitals and High School, 1381 work-charged personnel and 299 seasonal employees. On the electrical side there were 388 employees of different categories. The excavation of canals and water courses will also provide additional employment and income to the people of the Command Area.

(g) Pisciculture

The proposed cage pisciculture scheme in the Upper Kolab Reservoir will bring in new dimension in the income structure as well as food habits of the local people.

(h) Tourism

The Upper Kolab dam and the huge expanse of water in its reservoir will provide a place of attraction to the tourists. It is another prospective Vrindaban Gardan. All out efforts are required to be made to develop the spot as a tourist resort.

Drawing No.6 is the vicinity map of Upper Kolab Project which indicates the location of Upper Kolab Project site *vis-a-vis* its irrigation Command Area.

(i) The Other Side

All these prospects are the result, not without any stake. As large as 3171 families (14,405 persons) were uprooted of their ancestral abodes.

The standard of living, economic as well as social and cultural prosperity of those few families who have been resettled in colonies in an altogether new environment can be gauged only after irrigation water is available to them. No doubt the shifting of the displaced families to the settlement colonies has ended the isolation of the hill dwellers from the rest of the people in the district and their deprivation from adequate and necessary services. It brought them nearer to modern life. But there has been many a slip between the policy of rehabilitation as enunciated by the State Government and its implementation in the field. We shall discuss them in the following chapters.

(j) Peculiarities

The problem of rehabilitation with special reference to irrigation projects has certain peculiar characteristics. It is not a problem merely relating to mechanical or inanimate resources but rather a sensitive problem dealing with human beings which give rise to a number of social and sociological issues. The normal calculations of cost benefit here do not hold good. Cost of projects in terms of displacement of human beings can never be properly measured as these are social costs; but nevertheless they are very real. The problem becomes compounded by the fact that in many cases it touches the weakest sections of the population like the tribal population. It also involves special issues because, by and large, the benefits of these projects are rarely available to the people who are adversely affected by these projects. In other words, the sacrifice is from one group whereas, the advantages go to a completely different set of people. Hence there has been a lot of discussion on issues relating to rehabilitation of oustees of irrigation projects all over the world. Several activists organisations have raised their voice against the construction of large reservoirs. Before we discuss the lessons derived from Upper Kolab Project it may be worthwhile to know something about the reactions in respect of few other projects in the State as well as in the country.

Experience in Ghataprabha and Malaprabha Projects of Karnataka

In order to harness the water potential of Ghataprabha and Malaprabha rivers (tributaries of river Krishna) for providing irrigation facilities to the drought prone areas of Belgaum, Dharwar and Bijapur districts in Karnataka, two storage reservoirs were constructed, one across river Ghataprabha at Hidkal and the other across river Malaprabha at Naviletheertha. The former submerged about 6337 hectares of land affecting 22 village with a population of 15,660. The latter submerged 13,576 hectares of land affecting 43 villages with a population of 41,000. The reservoirs were impounded in 1972-73.

The submerged lands which were along the banks of river were fertile. Majority of the people were agriculturists. Though advance intimation was given about the submergence of the villages with the anticipated storage level of reservoir, people to be displaced were reluctant to shift because of the following main reasons:

(a) There was sentimental attachment to the place and houses, where they were born and brought up.

(b) Scale of assessment of land and property was not adequate.

(c) There was delay in payment of compensation.

(d) Compensation money paid to the oustees had been utilised otherwise, and they had no money to construct their houses in the new colony.

Experience in Srisailam Project in Andhra Pradesh

The Srisailam Project on river Krishna was completed in the year 1982. The water spread area of the reservoir is 237 sq. km. It submerged 85,000 acres of land affecting 78 villages and 22,000 families of 73,460 people. The oustees made their own rehabilitation arrangements at different places. However, the State Government provided the basic amenities such as drinking water, electricity, schools, etc. The State Government conducted a socio-economic study to examine the *(i)* pattern of utilisation of compensation money. *(ii)* changes in the socio-economic condition of the people and *(iii)* to suggest measures for the economic uplift for the deprived groups.

The study revealed the following :

(a) Of the total compensation money received by the oustee, 11 per cent to 14 per cent was used for acquiring land, 43 per cent to 50 per cent was invested on acquiring houses and about 15 per cent was spent on repayment of debts etc.

(b) There is a change in the occupational pattern. While the workers in agricultural sector declined, activities in non-farming sector increased. There was also decline in the live stock. In Kurnool district, 63 per cent of marginal farmers became agricultural labourers. Only 6 per cent of the agricultural household improved their economic condition. (*Source*-Rehabilitation Programme by S. Prabhakar).

(c) The study suggested that in order to improve the living condition of the displaced persons, irrigation facilities should be provided in the dry lands, fishing in the reservoir should be encouraged, employment generation programmes such as construction of roads in rural areas should be taken up and dairy development programmes should be encouraged.

Experience in Sardar Sarovar Project of Gujarat

The Sardar Sarovar Project has been taken up by the Gujarat Government with World Bank assistance, on river Narmada. This is perhaps the largest multipurpose project in the country aimed at irrigating 2.12 million hectares of C.C.A. and generating 1450 M.W. installed power at an estimated cost of Rs. 5793 crores at 1982-83 price. The proposed reservoir will submerge about 37,000 hectares of land of which about 11,000 hectares are forest land. It will also affect 248 villages involving 67,000 persons in Gujarat, Maharashtra and Madhya Pradesh. The Project in itself is an ambitious one and the rehabilitation action plan prepared for the oustees is equally ambitious. A lot of spade work has already been done to see that the oustees do not feel neglected. But without seeing the end result to come, a good number of Activist Organisations have raised their voices against the action plan so prepared for rehabilitation of the oustees. A good deal of deliberations have also been made in different forums about it and also about the environmental aspect. Rehabilitation of the oustees has to be done as per the Naramada Water Dispute Tribunal Award.

> In June 1991, the World Bank appointed Bradford Morse to organise an independent review of measures being taken to mitigate the human and environmental impacts of the Sardar Sarovar Projects. The report is already released. The following extracts from the Morse report need very close study:
>
> "The only resettlement policy applicable to all the three States is the Bank's. But Bank policy has not been respected. The Projects were not appraised in accordance with Bank requirements. Basic information had not been gathered and adequate plans for resettlement and rehabilitation were not in place."

"The history of environmental aspects of Sardar Sarovar is a history of non-compliance. There is no comprehensive impact statement. The nature and magnitude of environmental problems and solutions remain elusive. This feeds the controversy surrounding the Projects."

Experience in Subarnarekha Project of Bihar

The inter-State Project on river Subarnarekha will provide irrigation in 2,58,000 hectares C.C.A. (Bihar 1,60,000 + Orissa 93,000 + West Bengal 5,000) and moderate flood in Orissa and West Bengal. Possibility of generating about 50 M.W. power is also being explored. Two dams now under construction will submerge 25,613 hectares in Bihar and 4115 hectares in Orissa affecting about 60,000 persons. More than

20 per cent of them belong to Scheduled Tribe. Since it is not possible to provide agricultural land to each and every oustee family; besides allotment of available Government land, ceiling surplus land is proposed to be acquired for allotment. Further it is proposed to provide service opportunities in the project as well as in other Government Departments, self-employment schemes, vocational training, etc. The World Bank was financing this project until March 1989. Renewal is still under process. But as observed by the World Bank, there is much desired to be done in the Rehabilitation Plan now being acted upon by the project authorities.

Experience in Rengali Dam Project of Orissa

On the State front, Rengali Dam Project gives a very harrowing experience about the rehabilitation of its oustees. The dam was substantially completed in 1985 and one after another all the four number of units have since been commissioned. But payment of compensation for acquisition of land that was affected by the reservoir, displacement of people so affected and their resettlement are still unsettled issues. The Committee on Estimates, 1986-87 of the Orissa Legislative Assembly, on the Rehabilitation of displaced persons of Rengali Project in their 34th Report made the following observations/recommendations:-

(*a*) "cultivable lands given to displaced persons should be properly identified and possession should be given to the concerned person at the spot."

(*b*) A big tank constructed for irrigation purpose did not contain water, due to wrong selection of site.

(*c*) Employment opportunities should be created by implementing anti-poverty programmes.

(*d*) Land allotted for cultivation is neither levelled nor suitable for cultivation.

(*e*) Tanks and drinking water wells need further deepening and the tube wells, water of which contain iron should be properly treated.

The Special Committee on Rehabilitation constituted by the Orissa Legislative Assembly under the Chairmanship of the Hon'ble Speaker visited the Rengali Project area from 6.11.88 to 8.11.88 and observed as follows:-

(*i*) Reclamation of land has been done in a very careless and indifferent manner.

(*ii*) Family enumeration has not been done properly.

(*iii*) There are irregularities in payment of compensation money.

(*iv*) Lands identified for allotment were not plotted and demarcated correctly. Delivery of possession in the field was also not ensured.

(*v*) Land allotted has been encroached at several places resulting in disputes.

Man learns from follies. Hirakud, Balimela and Rengali are the three multipurpose projects in Orissa preceding Upper Kolab Project. Hence it cannot be said that the planners and executing personnel had to confront themselves with unknown and unexpected happenings while executing Upper Kolab Project. In that event there should have been either no problem or less of problem in tackling the resettlement and rehabilitation of oustees of this project. But what happened in reality ? Let us evaluate and see how for the rehabilitation policy has been implemented in the field. The experience in the field indicated several things. They can be categorised as (*a*) environmental, (*b*) Legal, (*c*) Socio-economic, (*d*) cultural, (*e*) human, (*f*) financial and (*g*) organisational. They are discussed seriatim in the pages following.

SECTION - B: ENVIRONMENTAL

Historical Perspective

From the very start of human habitation, man has been either working with nature or against her. In his effort to know the creator of Earth and his creation, he has been meddling with the environment without realising its consequences. In old Stone Age, he lived a nomadic life and earned his livelihood from hunting. Therefore, he moved from place to place in search of animals in dense forests. Thus, his life was very hard. For change, he also resorted to gathering of nuts and fruits alongwith hunting. Thus, he exploited the forest for his living. During the new Stone Age he began to learn many new things. The knowledge of agriculture brought epoch-making change to human life. He started to switch over from nomadic life to sedentary life. With settled life he started to improve the conditions of living. He developed better types of tools for more useful purposes and after the New Stone Age, built up great ancient civilization of Harappa, Mahenjodaro etc. All these great civilizations in the world rose on the river valleys. Later with the advent of machines, new sources of energy and new social, political and economic order he exploited his environment most with an objective of profit. Thus, with passage of time he marched from Forest Eco-system to space Eco-system; and the result has been disturbance only.

Water as a Resource

It is needless to emphasise the importance of water as a great natural resource. No life whether animal or plant can thrive without it. Agriculture, industry, energy, all place heavy demands on water. In this context, a river basin forms a distinct eco-system too. Man uses this eco-system for his benefits. But too much of use which can be termed as exploitation brings in hazards for him. In other words there is an environmental imbalance resulting in adverse effect on his own life as well as on the life of plant, bird and animal. Like any other natural resources, water resources have also limitations. Over-exploitation ends in imbalances. Resources of the Nile, Tigris and Euphrates, the Indus brought the best civilization, but over irrigation, water logging and salinity there also caused devastation.

UN Conference, 1972

The U.N.Conference on the Human Environment, held in

Stockholm in 1972, is considered to be the turning point regarding the linking of environment and development. After this conference, the UN General Assembly established United Nations Environmental Programme. At the above conference, M.F. Strong highlighted as under-

> "Man's capacity to destroy is, of course, most dramatically manifest in his possession of the technologies of mass destruction - in particular those of nuclear and biological warfare - but the relatively recent emergence of the environmental issues has revealed the more subtle but no less dangerous risks he faces from the uncontrolled use or misuse of natural resources and the technologies of production. It is an issue that transcends political boundaries and stage of economic development.
>
> Damaging conditions such as increasing salinity of fertile land, depletion of forest reserves, pollution of land and water supplies, maldistribution of populations, and the unanticipated and undesirable side effects of pesticides and fertilisers occur in a wide variety of countries, no matter what the state of their economy. Every nation is affected by pollution of the planet's atmosphere and of the oceans whether or not it contributes to that pollution. Thus, the subject of the human environment is global in character and of universal concern."
>
> (*Source* - "Parameters in the environmental impact assessment of Water Resources Projects"- an article presented by Dr.P.N.Kapoor, Civil Engineering Department, I.I.T. New Delhi in the Northern Regional Seminar on Water Resources Development - Environmental Issues and Solutions, 24-25 September 1987).

The aforesaid U.N. Conference opened the path for environmental planning and ecological impact assessment of any development project.

Global Forum in Environment and Development for Survival, 1988/1990

The Global Forum on Environment and Development for Survival, convened a Conference in oxford, England in April 1988. Later it organised a five-day conference on Environment and Development for Survival in Moscow in January 1990. About 200 parliamentary and spiritual leaders from all over the world met here and the conference was inaugurated by U.N. Secretary General, Javier Perez de Cuellar. Soviet President Mikhail Gorbachev, gave the valedictory address. Among others Dr.Karan Singh, then Ambassador of India in the U.S.A. participated in the deliberations. Dr.Singh made the following observations:-

> "The unprecedented human intervention into the environment has upset the delicate ecological balance that enabled Mother Earth--Bhavani vasundhara in the Indian tradition, Gaia in the Greek to survive for billions of years and become a unique crucible for the evolution of consciousness. Ruthless exploitation of non-renewable natural resources has created havoc and if allowed to continue can result in a series of major ecological disasters that can disrupt life on this planet."
>
> (*Source* - "Strengthening global consciousnes"-- article in the Sunday Hindusthan Times, Dt. January 21, 1990).

It is heartening that the human society has become aware about the ill effects of damage of environment and the Government have developed interest and responsibility in creating mass awareness for conservation and protection of environment. Workshops, Seminars, Orientation training programmes are organised to create such awareness. The Orisssa State Government have also instituted "Prakruti Mitra and "Prakruti Bandhu" awards. Mother "Earth Day" and "World Enviornmental Day" are also being observed.

Envirnomentalists Vrs Planners

For some years past, the environmentalists in our country have become very critical about the construction of Sardar Sarovar Project in Gujarat. At a particular point of time, with an intention to stall the World Bank financing for this project, several activists advanced their arguments against the project before the World Bank authorities in the U.S.A. For that matter the Bank appointed an independent review Committee under the leadership of Bradford Morse, formerly an Under Secretary General of the United Nations to review the measures being taken to mitigate the human and environmental impacts of the Sardar Sarovar Project. Morse has become very critical in his report about the environmental as well as Resettlement and Rehabilitation arrangements made by the Project authorities. The environmentalists are always against construction of large and multipurpose river valley projects. The critics term the environmentalists as conservationists or preservationists as the latter mainly aim at preventing the destruction of natural forests and wild life occuring as a direct consequence of the project. The following Table will give an idea about the magnitude of loss of forest wealth due to construction of some of the major projects in India.

S.No.	*Name of Project*	*Submergence of forest area. (in acre)*
1.	Sardar Sarovar (all three States)	26,476.00
2.	Tehri Dam (Uttar Pradesh)	3,952.00
3.	Rengali (Orissa)	85,545.25
4.	Upper Kolab (Orissa)	6,757.85
5.	Narmadasagar (Madhya Pradesh)	99,620.00
6.	Manibhadra (proposed) (Orissa)	16,500.00
7.	Hirakud (Orissa & M.P)	26,000.00
8.	Upper Indravati (Orissa)	6,100.00
	Total	2,70,951.10

It has been estimated that about 10 per cent of the forest cover in the country has lost its capacity of performing protective, productive and ecological functions, consequent upon initiation of as many as 250 projects all over the country since, 1951.

There is no second opinion that the loss of centuries old forest with its flora and fauna is an unmitigable tragedy which would concern anybody and everybody. Development activities demand a very heavy share on the "free-gift" of nature *i.e.,* forest. But with passage of time this gift remains no longer free. Mankind has already used up more this resource than Nature can reasonably provide or renew. This results in imbalance and if it is not corrected in time, may result in unforeseen consequences. Compensatory afforestation as contemplated by the Government of India under the Forest (Conservation) Act, 1980, is no substitute; because soil and water conservation system once lost, and non-regulation of climate during the interim period can hardly be compensated. The rapid decreasing rate of forest cover in Orissa is indeed alarming. While in 1978-79 it formed 43.46 per cent of the total geographical area of the State, in 1988-89 it came down to 35.82 per cent. The environmentalists have pointed out that out of total 32 crores and 90 lakh hectares of land surface in India, soil cover in respect of 12 crores and 90 lakh hectares have been completely lost. They have also warned that unless the rapid loss of soil cover is controlled, the country will face immense difficulties in meeting the food requirement of the growing population. There will also be difficulties in getting adequate grazing ground for the domestic animals. Hence it should be the endeavour of our planners not only to safeguard all our remaining forests but to substantially increase the area under forest.

Loss of forest wealth, flora and fauna are not the only causes worrying the environmentalists. They also point out a snag in the green revolution contemplated in the command area of the irrigation project. It is argued that high level of fertiliser and pesticides use, high yielding varieties of seeds and intensive irrigation will lead to untold problems. They are of the opinion that, "it is a technology that demands high and continuously rising inputs of chemical fertilisers and pesticides to maintain yields, but forgets that these critical inputs are petroleum derived products, based on a depleting stock of fossil fuels that may be exhausted in a generation. The Green Revolution pattern of agriculture is sustained by oil; what happens when the oil runs out?

In addition, depleting supplies of fossil, fuels also mean continuously rising prices of farm inputs which further fuel inflation. Worse, high levels of use of chemical fertilisers result in declining soil fertility over time and the eutrophication of water bodies, while pesticide use creates serious and well known health hazards—toxic contamination of food, high concentrations of DDT in mother's milk, carcing genia PCBs in human brain tissue.

(*Source* - The Sardar Sarovar Controversy - Are the critics right ? - An article published in the "Business India" October 30 to November 12,1989 Issue).

The Royal Thai Government, FAO and WHO organised the second Session of the Group of Developing Countries in Asia concerning Pesticide Residues in Thailand from April 2 to 5, 1986 and observed that (*a*) there are growing evidences of foods with excessive pesticide residues in the markets of developing countries and (*b*) there is urgent need for reduction of the said residue content of food stuff. (*Source* - UNEP News Letter—Volume 3, No.2, Bangkok, Thailand, July 1986).

The environmentalists also apprehend that construction of Tehri dam in Uttar Pradesh will obstruct the water flow of river Bhagirathi and consequently perennial Ganges may take different shape at the world famous pilgrime centres Hardwar and Risikesh.

Alternatives

The noted propounder of CHIPKO movement, Mr. Sundarlal Bahuguna, while addressing a Press Conference at Delhi on 20.1.90 observed that "instead of the storage dam proposed at Tehri, there should be a number of 'run-of-the river' schemes to produce the power from the flow of the river."

(*Source*-The Hindusthan Times, 21.1.90).

It is known in all quarters that the United States is the home of chemicalised agriculture. About 3 per cent of its population depend on agriculture and feed not only the entire nation, but also export a sizeable percentage of wheat, corn, sugar production. The author had got the opportunity to spend few weeks in the U.S.A. in 1988 in course of a training period on irrigation management and obtained first hand information that 10 per cent of the agricultural land there was set aside fallow with an intention to decrease farm output. Furthermore the farmers were given compensation for the said area in which they would have otherwise grown crops. What the author means to say is that mechanised as well as chemicalised agriculture in the U.S.A. has reached such high percentage of production that there is need for restricting it. Presently there is serious rethinking to do away with high degrees of chemical fertilisers, etc. The National Academy of Sciences as well as the U.S. Department of Agriculture have acknowledged that natural farming using modern techniques is a fully viable option. They have, therefore, recommended policy changes that would help transform U.S. agriculture in that direction.

Environmental Impact Assessment

As has been stated earlier, human interference with Nature, changes the environmental balance. Perhaps the most pronounced changes are effected by man's interference with water. The environmental impacts of river valley projects vary greatly with the physiography, climate, soil, social and economic conditions etc. This is precisely the reason for which Article 48 of the Directive Principles of the Indian Constitution lays down -

"The State shall endeavour to protect and improve the environment and to safeguard the forests and wild life of the country".

Article 51-A of the Constitution also speaks that "It shall be the duty of every citizen of India (g) to protect and improve the natural environment including forests, lakes, rivers and wild life and to have compensation for living creatures."

Government of India in the Ministry of Environment and

Forests, as a sequel to the decisions taken in Stockholm Conference of 1972 issued a set of guidelines prescribing Environmental impact assessment for all development projects; and more particularly in respect of water resources development projects. The primary objectives of environmental impact assessment of water resources projects are to ensure that;

(*a*) there is sustained development through optimisation of resources and

(*b*) there is adequate safeguard built into the project at the planning stage itself to eliminate altogether or at least minimise the adverse environmental impact.

When a dam is constructed across a river, a lot of changes take place. Most often these changes are not seriously thought about because of the immediate benefits that will acrue *viz.*, irrigation, power generation, flood control, development of pisciculture and tourism etc. But it should not be forgotten that there are many side effects which have got adverse impact and an assessment should be carried out to counter them and maintain a balance.

The guidelines for Environmental Impact Assessment of river valley projects was issued by the Government of India in January 1985 by the Ministry of Environment and Forest. Subsequently Government of India enacted the "Environment (Protection) Act, 1986 to protect, improve environment and to prevent the hazards to human beings, other living creatures, plants and property. It has been made mandatory that for each river valley project, environmental clearance is necessary and to that effect, detailed report containing proposal for treatment of catchment area should be furnished. Catchment area treatment should form an integral part of the river valley projects. The main objective of the project being increased agricultural production and rural development, soil conservation, afforestation, other anti-erosion measures, augmentation of rain water seepage, extension of the life of the project by reducing siltation and maintenance of ecological balance should be considered together. The cost of the ancillary items should be provided for by the concerned Department, but direct damage to the environment and ecology at the project site arising out of the project should be set right at the cost of the main project and the required funds for it should form part and parcel of the project.

Thus, the Government of India have stipulated that environmental clearance relating to the river valley projects should be obtained before the project is taken up for execution. But most often it has been seen that environmental clearance is given as a result of political pressure. On the other hand in the name of forest and environmental clearance, Government become too much fastidious in interpreting the legal provisions. Under Section 2 of the Forest (Conservation) Act, 1980, forest land refers to all categories of forests recorded as such in Government records. The position in the field may be different. But for utilisation of the said land which is even devoid of even a blade of grass; for non-forest use, clearance of Government of India is necessary. There is also many a slip between the identification of non-forest land and degraded forest land for compensatory afforestaton and physical implementation of the compensatory afforestation scheme.

The Upper Kolab Project was cleared by the Planning Commission and Government of India in 1976. Hence no Environmental Impact Assessment study was made in respect of this project. Nevertheless, preservation of environment was not neglected altogether in either planning or implementation of the project. The basic guidelines of environmental preservation were kept in view and they are discussed in three main areas as under:-

(*i*) Catchment are a (excluding submergence area).

(*ii*) Dam, submergence area and downstream river portion.

(*iii*) Command Area of the project.

Catchment Area

The catchment area intercepted at the dam site is 1630

sq.km. It is bounded by Lat 18° 23' and 18° 47' and 82° 32' 30" E and 83° 2' 30" E longitude and it looks like a square shaped plateau with an elevation of 850- 900 metre vide Drawing No.3. The area is a portion of the denuded Eastern ghats. Hence it is characterised by undulating table land profusely scattered with hundreds of little hills covered with low scrubs. Extensive "Podu" cultivation has made the situation still worse. It may be noted that this state of affairs has got no bearing on the construction of Upper Kolab Project; because the situation as on today has been continuing for last several years even before the project was contemplated. This is evident from the 1945 edition of Koraput district Gazetteer which is reproduced below:-

> "Near Koraput the hills have long since been denuded of all their forest growth, and are intermittently cultivated to their very summits with dry crops."

Catchment Area Treatment

The catchment area receives fairly good rainfall during monsoon. Further most of it comes during cyclonic storms resulting in peak flows in the river. There are no reservoir savings or lakes for intercepting the free flow from the catchment. On the other hand there are innumerable gullies. All types of erosion *viz.*, gully, sheet and rill have advanced too far in the area. Further high rainfall coupled with turbulent flow in the streams cause serious bank erosion. It has been estimated that the soil loss per year in the catchment is 17.14 million tonnes. (*Source*—Soil Conservation Scheme-prepared by the Directorate of Soil Conservation, Orissa for the catchment of Upper Kolab Project-1985-95).

Hence, massive soil conservation programme in catchment area is called for. But very negligible soil conservation work has been undertaken limited to few patches of cashewnut plantation. Before the Upper Kolab Project construction was taken up, some area was also treated with contour bunding. But they area almost non-existent now either due to erosion or buried under debris.

Area Requiring Treatment

The catchment area of Upper Kolab Project as in Drawing No.3 has been divided into seven watersheds namely (1) Guradi-1, (2) Kolab, (3) Guradi-2 (4) Galbeda, (5) Kerandi, (6) Kunduli and (7) Ganguli. They have been numbered in Drawing No.3 serially from (1) to (7). All the seven watersheds are spread over an area of 1,63,000 hectares. Of it 31,061 hectares alone come under submergence in Upper Kolab reservoir. 27,344 hectares are reserve forest and 20,566 hectares are village forest. The rest are other categories of land including 72,749 hectares of agricultural land. But the village forests can hardly be called forest, because they are completely devoid of any tree growth. Drawing No.3 also gives an idea about the forest cover in Upper Kolab catchment above the dam site. The pre-cambrian rock in the entire catchment area consists of quartzites, schists and granulite which are highly fractured, foliated and weathered. Therefore, the condition is very favourable to soil erosion. One of the most peculiar features of Koraput plateau is that four out of six numbers of major hydroelectric projects in Orissa namely Machkund, Balimela, Upper Kolab and Upper Indravati have been constructed on the rivers flowing from out of this plateau land. There are already visible and widespread effects of soil erosion in Machkund catchment and which have already caused anxiety. Any further deterioration in ecology of the area will not only affect the Upper Kolab reservoir alone, but will also have a serious impact on the entire hydropower generation system in the State. In this background, the entire area coming under the category of village forest and 72,749 hectares of agricultural land (thus, totalling 93,315 hectares out of total area of 1,63,000 hectares) need treatment intensively. For this the State Directorate of Soil Conservation has formulated a scheme costing Rs.423.14 lakhs to be implemented in three areas (Galbeda, Kerandi, Kunduli and part of Kolab) of the seven water sheds during a period of five years. The scheme mainly aims at plantation of cashew, sisal and coffee. The other soil conservation measures are gully control and field bunding. The Orissa Remote Sensing Application Centre, have prepared a scheme costing about Rs.25 crores on Environmental Management of Catchment of

Kolab Reservoir for getting external assistance from Norway. The scheme aims at mass scale plantation in the entire catchment area including the reservoir periphery, construction of check dams at the head of streams entering into the reservoir, construction of garland canal with silt traps in area 50 metres above high water line of the reservoir.

As has been stated earlier, the catchment area has been denuded to a great extent. Consequently many species of birds and animals which were responsible to a great extent in maintaining ecological balance in the area have disappeared. Hunting by the tribals inhabiting the area has further dwindled these species. Hence the aforesaid soil conservation schemes when implemented will go a long way in restoring the ecological balance.

Dam Environment

Temperature

No study has been made as yet in respect of the dam environment and that of the submerged area and the downstream stretch. The water spread area of the reservoir does

Year	*Maximum temperature (in O'C)*	*Mininum temperature (in O'C)*
1979	39.0	13.0
1980	-	12.5
1981	39.0	12.5
1982	44.00	13.0
1983	41.0	11.0
1984	41.5	12.0
1985	41.0	10.5
1986	40.0	3.5
1987	39.0	1.0
1988	42.0	12.0
1989 (till October)	40.5	13.5

(*Source* - Upper Kolab Project Office)

not appear to have influenced the temperature much. It would be evident from the above statement in respect of the dam site:-

So far as the river regime is concerned, there is also no change.

Stream Flow

Stream flow is very much dependent on the rainfall in the catchment area. In this regard, we have taken the rainfall statistics of seven number of rain-gauge stations located at Koraput, Jeypore, Pottangi, Padwa, Damsite, Nandapur and Lamptaput from 1976 *i.e.*, the year following the year of laying foundation stone for the project till 1989 (October).

Table 6.1 will indicate the details:

This may be precisely the reasons for which the reservoir has not filled up to its full capacity so far, excepting once. The Balimela reservoir gives a still harrowing picture. It has been having all along, a low water table consistently. The geographical plateau area for both the projects is one and the same.

Dr. R.C. Das, Chairman, State (Prevention and Control of Pollution) Board in the book "State of Environment in Orissa-I-Water Resources" 'Writes with reference to Hirakud Reservoir as follows:-

> "The decreased number of rainy days in a year is particularly disturbing, for it means monsoon to be more erratic now than before."

Dr. Das further makes a reference to the following observations of Dr.B.N.Sahu, retired Professor, Orissa University of Agriculture and Technology.

> "Hirakud dam has changed the path of monsoon making the rainfall in Padmapur Sub-division of Sambalpur district, the Patnagarh and Titlagarh Sub-divisions of Bolangir district very erratic. These areas have either drought or heavy rainfall."

Forest and Wild Life

The reservoir submerged 189.95 acres of reserve forest

Table 6.1
Total L Yearly Rainfall in MM

Year	*Rainfall recorded in raingauge stations*							
	State's actual rainfall	*Koraput*	*Jeypore*	*Pottangi*	*Padwa*	*Dam-site*	*Nan-dapur*	*Lampta-put*
1976	1012.90	935.6	1626.1	1077.4	N.A.	N.A.	1248.5	1547.7
1977	1326.90	1536.9	2209.0	1993.2	N.A.	N.A.	1743.3	2016.3
1978	1261.30	1233.3	2094.6	1454.7	N.A.	N.A.	1882.0	2182.1
1979	950.70	835.4	1598.0	1378.3	N.A.	N.A.	1165.6	1251.2
1980	1321.70	1519.1	2285.95	1619.5	N.A.	N.A.	1815.4	1969.4
1981	1187.40	909.2	1491.1	1327.7	N.A.	N.A.	1399.5	1640.0
1982	1179.90	1035.1	1466.3	1273.8	90.8.1	1760.01	1477.2	1407.17
1983	1374.10	1488.7	1791.8	1877.0	1440.3	1643.8	1570.4	1719.5
1984	1302.80	2128.5	2046.4	1340.0	1603.4	2129.22	1505.5	1304.0
1985	1606.80	1074.4	1641.1	1492.67	1283.25	1778.20	1525.4	1146.8
1986	1548.90	1783.1	1683.0	1661.7	1350.1	2237.0	1644.5	1426.2
1987	1040.40	1693.3	1680.5	1634.0	1198.75	1684.8	1356.5	1434.2
1988	1274.40	1572.0	1313.1	1337.0	1594.79	1363.2	960.5	931.8
1989	1272.20	1343.6	1471.3	1247.0	N.A.	1539.2	1210.8	940.3

The State annual average rainfall being 1482.20 mm, there were drought spells in the State in the years 1976, 1979, 1982, 1984 and 1987 to 1989. In the year 1986, the Upper Kolab Reservoir was impounded and that year the rainfall as recorded in the seven raingauge stations was very satisfactory. But it decreased in the subsequent years. The systema- tic discharge observations, taken by the project officers in river Kolab indicates that "gradually there is deteriorating trend in the total run off of the river."

(*Source*- History of Upper Kolab Project, Volume I, published by Government of Orissa, Irrigation & Power Department). The said source also indicates that "the number of rainy days are gradually reducing and there is greater dependence for annual run off on the storm run-off".

and 6567 acres of revenue forest besides other categories of land. The forest growth mainly consists of mixed deciduous forest species. But due to continuous practices of shifting cultivation, the land surface including the submerged area was practically devoid of vegetation. However, keeping the then existing forest rules in view, a sum of Rs.2,15,697.20 was paid to Forest Department of raising compensatory afforestation in lieu of the reserve forest area lost in the reservoir. Other categories of forest if lost were then not being considered for compensation.

There was not much of damage to the wild life as it was possibe for most of the wild life in the deforested area to safely shift to nearby forest areas which are comparatively dense. But it is a fact that not much work has been done to identify the loss of biological resources due to submergence. It is necessary that some inventory in this regard is done.

Fish Founa

No systematic study regarding the effect of Upper Kolab Dam on flora and fauna in general, and fish life in particular has been done so far. The river, before construction of the dam, was never notable for its fish wealth. However, the commonly noticed fish were Mahaseer (Barbus Spp), Leaches and cat fishes. Floating net pisciculture which has become very popular in the reservoirs of Indonesia, is being experimented on pilot basis in Upper Kolab reservoir. The result will be keenly watched for introducing it as a possible viable economic rehabilitation scheme for the oustees. We will discuss about this scheme in a separate heading later.

Mineral Deposits

No costly mineral deposits were identified in the area. Hence no such mineral deposits was submerged in the reservoir.

Archaeological Monuments

No archaeological monument was submerged in the reservoir.

Health Hazards

In tropical climate, normally the periphery area of reservoirs are attacked by malaria. Upper Kolab Reservoir is no exception. But this is not a phenomenon arising out of Upper Kolab Project. Koraput district gazetteer, 1945 speaks:-

> "As Jeypore, the original centre of administration, was found to be infested with Malaria the offices were moved in 1870 to Koraput, then only a tiny hamlet, in the belief that the place would be healthy on account of its height and open situation. But these expectations have not been fulfilled, and though Jeypore has now lived down its reputation for unhealthiness, Koraput is still dreaded for its fevers."

Thus, malaria fever is more than century old in Koraput area. The earliest of the experiments on this fever was in fact experimented in this tract. The district gazetteer speaks that-

"The Jeypore hill tracts have always been notorious for their malaria, and following on the discovery of the mosquito transmission of the disease by Sir Ronald Ross, some of the earliest investigations into malaria in India were conducted in the Koraput area in 1902 by Prof.J.W.W. Stephens and Colonel Sir Richard Christophers I.M.S., Major E.L.Perry I.M.S. spent sixteen months on malaria investigation in these hills in 1912-13 making Koraput his headquarters.

Before the formation of Orissa as a separate State, Koraput agency area including present Ganjam district was a part of Madras Presidency. In order to eradicate malaria from this area, then Madras Government had also taken several steps. As it is seen from the aforesaid district gazetteer as long back in 1930, "D.K. Visvanadhan of the Madras Public Health Department was appointed as special malaria officer for the Agency and began to tackle the problem by leading the streams underground through subsoil drains of loosely packed boulders covered with road metal and a layer of turf."

The National Malaria Eradication Programme is working actively in the State to eradicate malaria. Although there is nothing particular to Upper Kolab Project, the Indian Council of Medical Research are of the views that "Koraput district has stable malaria and transmission occurs almost throughout the year."

(*Source* - ICMR Bulletin, Vol.20, No.5, May 1990)

In the meeting of the Central Council of Health and Family Welfare held on 1-3 February, 1989, it was resolved that the State Governments " should have a high level interdepartmental committee to screen proposal involving major irrigation and development projects and ensure that the project authorities incorporate anti-malaria components in the project during the planning, construction and in the maintenance stage. The engineers should be trained in anti-malaria operations."

Seismicity

According to the records of the project authorities, no earthquake of any significance had occurred in the past in the project area. There is also neither any fault nor any fracture zones known in the vicinity of the dam. But Dr.N.C. Malik in his essay "Spillway for Satiguda Dam" (Orissa Engineering Congress Souvenir 33,47-55) writes that there was felt a mild earthquake on Sunday the 1st May 1988 for seven second near Satiguda pond which has been created down below Upper Kolab Dam to catch the tail race discharge of U.K. Power House. His observations are reproduced below:-

> "Satinala descends through a series of rapids and small falls from Koraput plateau to Jeypore plateau. The 1138 Metre long Satiguda Earth Dam spans two hillocks on either side of Satinala near village Badaliguda, 3 Km above Jeypore town and 15 Km below Kolab Power House.
>
> From the drill data of spill way it was found that the rock underneath was greiss charnochite with moderate number of joints in the left side and massive in nature in left flank.
>
> In a small patch under the right training wall of the spillway the greissic granite was found to be impregnated

by pink colour granite. The joints are thin, some times strained but free from extraneous materials."

Against this seismic and geological back drop, scientists can very well assess the pressure and strain of Upper Kolab reservoir water mass on the bed rock to cause any anxiety or not.

Sedimentation

Due to heavy soil erosion in the catchment, silt load of the reservoir is very heavy. The dead storage capacity provided in the project takes into account the silt load. The engineers advocate that there is no possibility of flood in the upstream, and the aggradation will not create serious problem in the upstream and the capacity of the reservoir. But it is no gainsaying that all the seven watershed areas as depicted in Drawing No.3 need careful and special environmental protection programmes. The reservoir periphery land which are moderately steep are apprehended to be creating sheet wash and carriage of silt. The engineers have calculated that the actual silt load at the dam site is 0.70 MM annually. On the other hand the State Soil Conservation Directorate have assessed the silt loss in the catchment to be 17 million tonnes annually. But all said and done, rate of siltation in the reservoir has to be checked.

Catchment area treatment *i.e.*, afforestation and soild conservation measures are advocated to be the most convenient and successful measures to prevent/reduce silt deposits in the reservoir. No systematic study has been made so far to gauge the impact of these measures on siltation. But it can hardly be denied that a denuded land is more prone to erosion of various nature than a covered land. Hence the proposed soil conservation measures under catchment area treatment will go a long way in arresting silt in the reservoir. Urgent steps should be taken to implement the schemes prepared in this regard.

Command Area

Loss of Forest Wealth

Besides the loss of forest due to submergence in the reservoir area, a sizeable forest land of different categories (given in Table 6.2) in the Command Area was also reclaimed for enabling the project authorities to set up resettlement colonies. The loss has not been compensated so far. Among several ill impacts on environment due to this loss of forest wealth, the most noticeable are the disappearance of the source of fuel wood and pasture in the area.

Ground Water Regime

From the observations of ground water in the ayacut, it is seen that ground water table in much below the ground level. The small change brought about by irrigation water will not create any problem for water logging because of the fact that the ground slope is steep enough for drainage However, effective drainage measures are necessary to be taken while doing micro-planning of the ayacut.

Water Quality

The water of river Kolab down the dam site, after due test has been found to be suitable for domestic as well as industrial use. But the pollution due to effluents of SEWA Paper Mill near Jeypore has to be guarded.

Suggestions

Construction of a Project of this magnitude is likely to have considerable impact on the surrounding eco-system. Hence at the project formulation stage, a systematic, Scientific impact assessment study should have been undertaken. Unfortunately this has not been done. Although no serious environmental hazard has resulted so far, much is desired to be done in protecting the environment, especially in the catchment area. The greatest adverse situation which would worry anybody is the rainfall pattern. There has been noticed radical change in the quantum of rainfall after 1986. It is on the lower side compared to previous years. Out of 1482.2 MM mean annual rainfall of the State, 85 per cent occurs during the four monsoon months. That too it is indefinite, resulting in flood in one year, and drought in another.

(a) Soil Consevation Measures

Experience indicates that oustees in reservoir projects always prefer to move into the Upper catchment belt immediately surrounding the reservoir in spite of the fact that the downstream Command Area holds better propsects. It happens more so in respect of tribals who have a special preference to stay in uplands. It is learnt that in the instant case, a sizeable oustee population moved into the Upper catchment belt immediately surrounding the Upper Kolab Reservior because of the hilly terrrain. But the influx of oustees must have put further pressure on the economy of the area. Traditionally most of these people were practising "Podu" or shifting cultivation. Hence shifting cultivation in the areas of domicile is apt to be intensified causing thereby further impact on the environment. Launching of large scale soil conservation measures and particularly afforestation measures in the catchment area of the reservoir seems to be the most viable answer. It is urgent because the Eastern ghats to which the Koraput Plateau belongs is one of the important geographical features of the Indian Sub-continent that plays a vital role in governing the climate of Orissa. The plateau has been degraded too much, resulting in serious environmental imbalances.

(b) Propagation of Plant and Animals

So far as loss of biological resources is concerned, not much work has been done to identify them. It is necessary that some inventory is made in the submergence area. Plant and animal species on the brink of extinction should be identified and method of their propagation should be found out. It should be ensured that a perfect eco-system is created to accommodate nesting birds and migration of wild life to sanctuary areas.

(c) Prevention of Malaria

It is not late to identify the already infested and probable malaria infested pockets, if any, both in the reservoir area and Command Area and take intensive anti-malaria measures on priority. It may be noted that acquatic weeds in canals and water courses are likely to cause serious problems for efficient flow of irrigation water. Consequently they may be the breeding grounds of mosquitoes. The problems are likely to be aggravated as the canals are not being lined. To overcome this problem it is necessay that water distribution may be on rotational basis, so that there will be some off period for the canals to drain out water.

(d) Soil Erosion in Command Area

The soil quality with 1:1 canal embankment slope leads to slides especially during monsoon rains. Mitigating measures which include careful planning of borrow pits, plantation should be taken.

(e) Soil fertility

The irrigation project provides for 186 per cent cropping intensity. This is likely to have negative impact on soil fertility. Sufficient organic manure may be used to neutralise this.

(f) Drainage and Conjunctive Use

It is necessary to plan and develop and utilise ground water in conjunction with canal water. It will also help in controlling probable rise in subsoil water level. Adequate drainage facility should be there to drain away excess surface flow in fields.

Section - C : LEGAL

Acquisition of Land and Payment of Compensation

Like many other States, the State of Orissa has got a legal frame work for acquisition of land for Government purposes, evaluation of the property so acquired and payment of compensation for the same. We have already discussed about the procedure adopted in the acquisition of land and payment of compensation money after necessary valuation of property. The following are the lessons learnt in this regard:

Valuation of Land

Market value of land to be acquired was calculated after taking into account, the actual sale deeds in the locality within three years prior to 4 (1) Notification under Land Acquisition Act, 1894 plus additional market value at the rate of 12 per cent per year for three years and 30 per cent extra towards solatium charges. But the irony is that value of land in areas to be submerged in the reservoir gets depressed from the moment it is known that the project would be taken up in that area. There is a time gap between the knowledge about proposed project and actual acquisition of land. Secondly in order to avoid payment of stamp expenses, and registration fees the land sellers undervalue their property. The third factor which weighs consideration is that many unscrupulous middlemen and buyers persuade the innocent simple natured tribals to quote a lower rate than the real value. Fourthly the Orissa Regulation II of 1956 which protects the interests of tribals and S.C. make their land inalienable to people belonging to other communities. More than 50 per cent of the oustees belong to S.T. and S.C. Hence not much sale figures are available to help assess a reasonable price for the land acquired. It is, therefore, necessary to streamline the valuation of landed property so that the displaced person does not face hardship. This can perhaps be achieved if transactions made in areas in the close proximity of the submergence area form the basis in fixing the basic price and not the sale price as recorded in the sale registers in the office of Sub-Registrar.

Valuation of Trees Standing on the Land

In Upper Kolab Project, valuation of trees was made basing either on cost of fruits or on cost of wood whichever is higher. In case of unspecified trees, the cost was calculated taking them as firewood. This procedure does not seem to be a scientific one.

In case of fruit bearing and timber trees, which have definite market value, compensation money could be fixed keeping the market value of the wood or a certain number of years' purchase of the annual value of the produce whichever is greater. For this, ten years' time could be the ideal period. It is hence necessary to make amendment to the existing procedure.

Payment of Compensation

Annexure II gives the details of compensation money paid for the reservoir area. Table 6.2 would indicated that there is a wide gap between the time institution of land acquisition cases and the actual date of payment of compensation money. As many as 47 cases instituted during the period from 1978 to 1985 and involving compensation amount of Rs. 1.56 crores were still pending on 20.1.1987 for payment due to some reason or other. There was even delay between admission of bill and encashment of the same. Thus, in the interim period, the land value would have increased. There were also no buyers to come forward for better price as the issue of notification under section 4 (1) of the L.A. Act made everybody apprehensive about the submergence. The oustees thus passed through a period of uncertanity.

This lapse in delayed payment was accepted as there is no statutory provision to limit the delay excepting provision for payment of interest money.

ENUMERATION OF FAMILY

(a) Date of Reference

In Upper Kolab Project, enumeration of family has been done taking the date of laying foundation stone (dt. 11.6.1975) for the project into consideration without updating at any subsequent time. The State Government amended this provision in their Resolution No. 58354 dt. 23.12.87 and laid down that date of publication of 4 (1) Notification under L.A. Act, 1894 will be the reference date for the purpose of enumeration of families. This amendment did not have retrospective effect. Consequently several persons who otherwise would

Table - J
List of L.A. Cases pending for payment of compensation
(Para - 6.2.13)

Sl no	*Case No & Year of ins-titution.*	*Name of village*	*Area (in Acre to be acq-uired.*	*Bill No.*	*Date of Bill*	*Amount of Compen sation.*
1	2	3	4	5	6	7
1.	54/78	Sindhiput	94.12	112	8.5.86	68,920.15
2.	94/78	Dumuriput	144.21	113	“	1,70,047.10
3.	98/78	Pitaguda	80.41	129	“	49,244.00
4.	107/78	Chandalamunda	26.28	119	“	88,286.15
5.	25/79	Lanjimuku	0.80	126	“	4,922.00
6.	26/79	Janiguda	2.14	111	“	7,478.50
7.	28/79	Kumariput	56.76	106	“	2,20,987.25
8.	32/79	Kingeiput	0.26	134	“	1,196,20
9.	36/79	Badamunda	0.04	136	“	187.75
10.	50/79	Khilaput	38.99	132	“	1,03,642.15
11.	55/79	Charangaon	117.04	114	“	4,50,469.70
12.	61/79	Bijaput	0.65	118	“	1,548.35
13.	62/79	Machaput	5.82	114	“	24,390.25
14.	67/79	Charanguli	151.41	115	“	8,65,831.50
15.	63/79	Dusura	0.13	173	13.5.86	415.93
16.	4/79	Masingput	84.13	493	17.11.86	81.60
17.	29/79	Umbel	86.19	497	“	4,06,449.15
18.	49/79	Chhagaon	371.12	449	“	23,03,453.30
19.	64/79	Bodlipondi	64.16	494	“	3,22,492.40
20.	8/80	Khalpadi	154.67	102	8.5.86	6,47,726.40
21.	10/80	Jogiput	0.42	138	“	874.20
22.	13/80	Kukudaput	85.75	133	“4,87,366.05	
23.	14/80	Kanheiput	0.41	109	“	1,566.00
24.	18/80	Parajakhubudi	141.36	149	“	3,02,172.65
25.	19/80	Parata	2.00	116	8.5.86	13,104.25
26.	20/80	Jantaput	119.80	140	“	3,09,525.80
27.	23/80	Debatahanjar	14.75	120	“	29,531.85
28.	25/80	Kadrimaliput	53.19	148	“	3,43,501.55
29.	32/80	Gopalput	15.41	139	“	1,00,957.95
30.	35/80	Mahipalput	7.30	121	“	26,264.90
31.	36/80	Limaput	32.58	135	“	60,902.60
32.	31/80	Kotagaudaput	0.18	174	13.5.86	236.50
33.	34/80	Mondieput	1.59	175	“	11,397.65
34.	2/80	Badatankua	274.36	498	17.11.86	14,78,948.00
35.	5/80	Patrapondi	38.34	495	“	1,61,298.00
36.	9/80	Arna	364.70	500	“	13,855,050.00
37.	12/80	Chandalamunda	338.93	501	“	14,49,638.45
38.	16/80	Komel	1.51	491	“	3,251.15
39.	17/80	Gumelput	76.92	496	“	1,89,982.35

(Contd.)

Table 6.2 (Contd.)

1	2	3	4	5	6	7
40.	30/80	Thusuba	50.41	492	17.11.86	3,09,836.80
41.	8/81	Badapada	185.65	117	8.5.86	8,41,133.35
42.	10/81	Sahuput	86.64	131	"	4,82,978.65
43.	12/81	Kutugam	0.52	110	"	2,887.35
44.	15/81	Gulpondi	103.91	503	17.11.86	4,49,934.75
45.	29/82	Chikapur	72.54	343	22.8.86	3,43,133.00
46.	27/83	Kumariput	1.15	N.A.	N.A.	2,918.00
47.	2/85	Parajapondi	621.66	"	"	11,41,605.00
			4172.05			1,56,70,071.00

(Besides nine number of cases were pending at different levels for publication under the Act and one case was pending at the Government level for sanction of estimate).

have been eligible for enumeration as families did not get this advantage. Out of total 3171 number of displaced families, 199 number of families were yet to be displaced as on 23.12.87. Further out of them, 104 families were shifted very late in 1991 only. The figure of displaced families which was arrived at by taking the date of laying the foundation stone of the project without updating, is anomalous. There would have been increase in the number of families meriting consideration for receipt of rehabilitation facilities had dates of publication of notice under section 4(1) of L.A. Act been taken into account and the figures were updated from time to time. Of course, any tampering at this point of time may create complications.

Further there was a gap not only between the date of issue of notification U/S 4 (1) of L.A. Act and date of payment of compensation money, but also between the date of payment of compensation money and the date of physical shifting of the displaced families. But no updating of the list of families was made at any subsequent time which placed the oustees in disadvantage. Fortunately this lapse has been made good in the modifications brought out in Irrigation Department Resolution No. 31888 dt. 21.8.90. Without becoming very stringent towards these 104 number of families who were shifted in 1991; the project authorities should have ensured that re-enumeration was done at the time of physical shifting of this particular group of families.

(b) Age Factor

Another lapse committed in the matter of enumeration is that although in Irrigation and Power Department Letter No 43782 dt. 25.10.89 it was clarified that major and married sons who are of and above the age of 21 years will be treated as families for the purpose of rehabilitation, the same was not made applicable to Upper Kolab Project, although the fourth phase of displacement was still to be implemented then. Age limit has been further relaxed to 18 years irrespective of the fact that he is married or not. This issue should have been taken into account while shifting 104 families.

Land for the Landless Agricultural Labourers

Under the Orissa Government Land Settlement (Amendment) Act, 1989 "landless agricultural labourer" means a person who has no means of livelihood other than agriculture, provided -

(*a*) "he or any member of his family owns no land excluding homestead, and

(*b*) his total annual income, together with the total income of all the members of his family who are living with him in common mess from all sources, does not exceed Rs. 3600 or an amount which the State Government may, by notification from time to time, specify in that behalf. In this regard "person" means any person, "the total extent of whose land excluding homestead, together with lands held by all the members of his family who are living with him in common mess is less than one standard acre and whose total annual income together with the total annual income of all the members of his family living with him in common mess, does not exceed Rs.3600 or an amount which the State Government may by notification from time to time, specify in that behalf."

On the other hand under the Orissa Government Land Settlement Rules, 1983, a "homesteadless person" has been defined as a person who together with all the members of his family living in common mess; - (*i*) does not have an annual income together with the annual income of all the members of his family from all sources exceeding Rs. 3600 or any amount that may be fixed by the State Government from time to time, (*ii*) does not have any homestead land anywhere in the State and (*iii*) owns less than one standard acre of agricultural land.

The aforesaid Act and Rules further provide that landless agricultural labourer will be settled with land upto one standard acre or to the extent to which the land already owned falls short of the standard acre. Further with a view to providing shelter to all who are homesteadless, not exceeding five decimals of homestead land is to be allotted. Of course, all these provisions are applicable subject to availability of Government land.

In case of Upper Kolab Project, out of 3171 number of families displaced, only 528 number of families were accommodated in the project sponsored colonies; the rest having opted to make their own resettlement arrangements.

There has not been any monitoring about the said resettlement progress. But the question remains to be answered; whether those 2643 families who refused to go to project sponsored colonies would be treated as homesteadless or not ? The author's answer is in the affirmative. The moment the person is shifted from the submerged area is rendered landless as well as homesteadless provided he does not possess land and house elsewhere other than in area of displacement. Thus, he is in a disadvantageous position and has got no fixed source of income; provided there is nobody in his family who is serving in any organisation or trading on any profession outside the submerged area. It is, therefore, very much necessary that legal provision is made so as to enable such of the oustees to avail of the same. To avoid complications, certificates should be issued to those oustees in token of the fact that they are oustees, and no land under the rehabilitation policy could be allotted to them by the project authorities. Oustees refusing to accept land so allotted to them and or opting to make their own arrangements will be deleted from this category of oustees meriting consideration for issue of aforesaid certificates. The certificates so issued can be utilised for allotment of homestead land in the new area of resettlement, provided the oustee has got no house/homestead land elsewhere.

Compensation for "Paramboke" Land

The Madras Estates Land Act, was applicable to Koraput district, before the revenue laws of Orissa came into force there. Under the said Act status of "Gram Kantha Paramboke" land was being regulated through executive instructions only. These lands were treated as communal land in spite of the fact that individuals settled thereon after constructing their houses and enjoyed the usufructs of the trees standing on the said land. The State Government in Revenue Department letter No. 1951 R dt. 8.1.1975 issued instructions that payment of compensation in respect of "Paramboke" lands where such lands are required under L.A. Act should be made. It was further clarified in Revenue Department letter No. 42397 R dt. 1.7.83 to the Secretary, Board of Revenue, Orissa that occupiers of "Paramboke" land have got transferable right in respect of homestead land.

In the above context, the displaced persons who had occupied "Paramboke" land, ought to have got compensation money under the L.A. Act. But assuming that the occupiers had got no private rights accured in respect of "Paramboke" land, the authorities did not make any provision for payment of compensation towards acquisition of "Paramboke" land. It is a serious lapse which cannot be made good.

Allotment of Land to Displaced Persons

According to the provisions existing then, the oustees family who opted to settle down in the project sponsored colonies were allotted with land at the rate of 6 acres of reclaimed unirrigable or 3 acres of reclaimed irrigable agricultural land free of salami and 0.5 acres of homestead land irrespective of the fact that they are landless or possessing land upto 6 acres. The oustee family was also at liberty to get part land part money in cash for the remaining part of the land. State Government in Irrigation and Power Department Resolution No. 240 dt. 3.1.89 decided that the aforesaid ceiling will be reduced to 2.5 acres of irrigable land or 5.00 acres of unirrigable land. There was no change in allotment of 0.5 acre of homestead land. The amendment was done with a view to extending the benefits of "small" and "marginal" farmers to these oustees. In the said resolution and Resolution No.8554 dt. 28.2.89 the ceiling of rehabilitation cash grant was also enhanced from Rs. 14,040 per family to Rs. 17,875 and Rs. 20,075/- per family respectively. (There was been further amendment in August 1990). These resolutions were made applicable only to those projects where the rehabilitation of displaced persons had not started by then. Strictly speaking, the oustees of Upper Kolab Project are not to get the benefits under the amended provisions. But some unesiness lies in the minds of 104 families, who shifted in 1991.

Suggestion

(a) Allotment of Private Land

In compelling circumstances, the people from the submerged area have been displaced. It is argued that those oustees who refused to move to the project sponsored colonies and shifted to their places of choice are not entitled to get land at their new places under "land for landless" scheme of Government. But if the land offered in the colonies were refused, it is hardly surprising that the people rejected the same, having no good soil cover, no irrigation, no adequate drinking water facility, etc. Hence where Government fails to allot suitable repeat suitable land in sponsored colonies, efforts should be made either to acquire private land or purchase them for the oustees in the Command Area and neighbouring areas. On failure to do so, oustees resettling in other places on their own, may be treated as landless/home steadless and benefits of Orissa Government. Land Settlement Act may be extended to them. This is subject to the condition that these oustees have not got any land to settle down, or any other fixed source of income to fall back upon.

(b) Enactment for Resettlement

Like the "Maharashtra Resettlement of Project Displaced Persons Act, 1976" some enactment may be made to provide for the acquisition of private land in the Command Area for the resettlement of persons displaced from lands which are acquired for projects of public utility.

(c) Valuation of Land

Instead of determining the value of land basing upon sale documents which are not true representative sales, it may be done by capitulising annual net profit from the land to be acquired.

SECTION D - SOCIO-ECONOMIC

Introduction

Involuntary resettlement is always a tougher proposition than voluntary resettlement. While in the case of the latter, the oustee is compelled to make his own arrangements for shifting and resettlement; in the case of the former, resettlement of the oustees becomes the responsibility of the Government. The proverbial saying (Poetry, women and creeper cannot thrive without shelter. Literature prospers due to the patronage of lovers of literature, woman is dependent on her husband and creeper takes shelter in trees) is applicable to the oustees. This is more so in a country like India which believes in socialistic pattern of society. The large scale displacement of the people from their ancestral dwellings because of submergence in the reservoir, is always a traumatic experience. It leads not only to economic and social insecurity but also to emotional disturbance. The people before displacement used to earn their living through some vocation, business or labour. They all had their long and short term economic plans keeping their resources and requirements in view. They were fully conversant with the people and society around them and depended on their friends and neighbours in any contingency. The environment and culture in which they lived were very fami- liar. Hence it is very natural to become apprehensive about the new environment at the place of resettlement.

Baseline Survey

It is against this background that the Government machinery entrusted with the responsibility of resettling the displaced persons make all sorts of bright promises and the displaced person expects that the facilities provided for resettling him at the new site would ensure better living standards than he lost at the old habitation. It is, therefore, very much necessary that a socio-economic survey in the submergence area or affected area is conducted so that the resettlement measures to be planned provide for the social, cultural and economic needs of the families displaced. The State Government Resolution No. 18473/78 dt. the 20th May, 1978, in Irrigation and Power Department clearly stipulates that, "no submergence will be allowed till rehabilitation facilities are established according to the schedule of submersion. For sometime past, Orissa has been getting finances from the World Bank for construction of several major and medium irrigation projects. Presently Upper Indravati Hydro-power Project is under construction with Bank assistance. Subarnarekha Irrigation Project and few other projects are also in the pipeline. The Bank does not give financial assistance either for land acquisition or for resettlement and rehabilitation of the oustees. But the project agreement executed with the Bank requires that an action plan for resettlement and an implementation schedule should be completed two years before the project causes submergence of land. Apprehension of Bank's withdrawal to assist in the construction of the project due to nonfulfilment of the terms of agreement is always there. That apart, it is the responsibility of the State Government to take effective steps and create a viable resettlement and rehabilitation plan and to implement it sufficiently before the reservoir begins to fill.

Condition Prior to Submergence

It is unfortunate that in case of Upper Kolab Project, no socio-economic baseline survey was undertaken. The bench mark survey done to assess the area of submergence was also neither systematic nor perfect . This is the reason for which

F.R.L. 858 M initially demarcated in the field was subsequently found to be incorrect. It was actually 856 M. This necesistated fresh survey resulting in identification of 104 additional number of families for displacement at a very late stage.

The rehabilitation policy of the State Government also does not speak anything about the arrangements to be made for the people residing in villages which become marooned consequent upon impounding of water in the reservoir. One such village is "Paroja Pondi" which was detected to be marooned on three sides resulting thereby in disconnecting the communication system. Hence as per the recommendations of the Rehabilitation Advisory Committee, all the 227 families residing in the said village were shifted. Further survey conducted in 1985-86, after construction of the dam was completed to identify additional marooned villages, if any. The Rehabilitation Advisory Committee in its 9th meeting in February 1987 discussed the problems of marooned villages and recommended the following:-

(*a*) Several families in village 'Chikapur' and 'Daraput' should be shifted as alternative road communication was not possible.

(*b*) Link roads should be constructed in case of 'Tiadiput' and 'Podapalma' to ensure effective communication system. In case such construction is assessed to be more expensive than rehabilitation cost, the people from these villages may be shifted.

(*c*) Where road construction is not possible because of reservoir formation, boats should be provided at the project cost to provide communication link with the market places. For that matter three ferry ghats near 'Dangadeule', 'Mundagarh' and 'Bhitarchenda' villages were identified.

Government accepted all these recommendations.

Works relating to land acquisition and rehabilitation started in right earnest from 1981-82, and the reservoir started filling in from 1985-86. But survey of the area coming under submergence was not completed until June 1989.

Population Affected

The following is the abstract of the number of villages affected and number of families displaced due to construction of Upper Kolab Project. It is not the practice in Orissa to keep account of number of persons belonging to these families. However, the author collected the population statistics too. Details are in Annexure-IV.

Phase	*No. of Villages affected.*	*No. of families affected.* S.C.	S.T.	Gen	Total	*Total population displaced*
1st	5	129	431	135	695	3,058
2nd	21	163	569	678	1410	6,422
3rd	23*	90	375	323	788	3,550
4th	9*	15	27	52	94	448
5th (Recovery)	2*	-	4	100	104	516
Satiguda Pond	2	45	15	20	80	411
Total	62	442	1421	1308	3,171	14,405

* 2 + 1 + 2 = 5 are repeat villages.

Thus, S.T. people formed about 45 per cent and S.C. people formed about 14 per cent of the total families displaced.

Pre-submergence Socio-economic Status

As has been stated earlier, no socio-economic base line survey was conducted in this Project. The author interviewed some of the settlers in November 1989 and collected information in a format (in Oriya) developed by himself from the 424 families resettled by then in colonies. An English copy of the sample format is reproduced in Annexure-VI. He also devised format Annexure-VII to collect informations on the submerged villages. It is also an English version of the format in Oriya. It revealed that most of them were agricultural labourers with a very meagre income. A very small group of oustees from the bottom of the economic heirarchy was represented by small and marginal farmers who used to cultivate their own land. During off season some of them used to work as labourers away from their homes on daily wage. Those who could not get any job collected forest products and sold them at Koraput, Sunabeda and Jeypore. There were among them Carpenters, black smiths, Potters in small strength. The main crops grown were short duration paddy, ragi and niger.

The residential houses were of either mud walls or rids, "ballahs" and bamboos plastered with mud puddle. Roof was mostly thatched by paddy rids, date plam leaves, etc. Villages were not of any set pattern. Houses were set apart in a scattered manner as the occupants liked. Roads within the village were not planned. For drinking water, the people were mainly dependent on river and streams nearby. Those who were away from the rivers and streams, were dependent on the open wells in the village area. The villages were not well connected with market centres. Foot tracks along hill slopes provided communication. Bullock carts which were very few in number used to ply in dry season. The engineering survey party of the project, however, used jeeps in course of their routine rounds.

The population was mostly dependent upon agriculture either directly or indirectly. But fertility of the land submerged was not good as most of them were 'dangar' land in classification; excepting small patches of 'jhola' land here and there in the villages and by the side of the streams. On the average, the cost of agricultural land was Rs.2500 only per acre. Statistics collected in respect of 3000 displaced families from the project office give the following information:-

(i)	Landless	...	840	families
(ii)	Having about one acre of land	...	281	families
(iii)	Having one to two acres of land	...	371	families
(iv)	Having two to three acres of land	...	336	families
(v)	Having three to four acres of land	...	234	families
(vi)	Having four to five acres of land	...	170	families
(vii)	Having five to six acres of land	...	134	families
(viii)	Having more than six acres of land	...	634	families
	Total		3000	families

Thus, as many as 28 per cent of the total families displaced were landless. Most of the remaining families were also not in a position to make their both ends meet because of low fertility of soil. Collection of firewood for sale in the nearby town areas was a convenient source of subsidiary income. The upsurge of construction activities in Upper Kolab Project, NALCO at Damanjodi, road development, etc., took many to wage earning tool. The following statistics would give an idea about occupation-wise break-up of the 3000 displaced families:-

(*i*)	Dependent purely on agriculture	1160	families
(*ii*)	Dependent on service	87	families
(*iii*)	Dependent on business and trade	9	families
(*iv*)	Dependent on wage	1744	families
	Total	3,000	families

Agricultural implements were of crude kind. The plough was made of wood and iron. Other implements in common use were the levelling board, soil scrapper, harrow and the common sickle. Hill side slopes were cultivated by means of hand mattocks (Kodi) and forks.

Most of the people believed in Hindu religion. Only a very few people were muslims. The tribals maintained their separate entity. They believed that diseases are due to 'divine' vengeance for wrong doing, effects of 'evil spirit', etc. and the treatment methods were resorted without any scientific and medical bearing. Only when they did not get relief from these measures, and if transportation could be afforded they used to go to P.H.C. The tribals belonged either to 'Paroja' or 'Gadoba' community. They were less advanced and they cared little for their individual rights. They were fond of remaining in groups and were accustomed to do only minimum of work that will bring food sufficient for the day. However, after coming in contact with the urban people in course of visiting market places, they had started showing interest in hoarding for the future. Most of them were used to country liquor prepared at home out of rice. "Salap" trees were taken in high esteem, as they used to provide liquor too.

Reaction of Oustees to Shift to Colonies

Initially it was estimated that about 6,400 number of families will be affected by Upper Kolab reservoir in 147 villages and about 25 per cent of them may go to the project sponsored colonies to be set up in Boriguma, Kotpad and Sashahandi areas where 12,700.84 acres of land as under had been identified:

(*i*)	Batasana and Nuagaon Circle in 23 villages	9,432.00	acres
(*ii*)	Boriguma Circle in 25 villages	3, 248.00	acres
	Total	12,680.00	acres

In the first phase it was decided to confine the rehabilitation activities to batasana and Nuagaon Circle under Kotapad Tehsil. Accordingly reclamation work of land was taken up in that area to set up seven number of colonies; six numbers in Batasana Circle and one in Nuagaon Circle. Reclamation work in Boriguma Circle was deferred. As a matter of fact no reclamation work was taken up there because of poor response from the oustees to shift to colonies. Hence out of 9,432 acres of land in Batasana and Nuagaon Circle, 9425 acres of land was made available after due reclamation, to the Resettlement and Rehabilitation Organisation of the Project. The colony-wise land particular are furnished in Table 6.3.

Out of 3,171 families in all, in 60 villages and two hamlets displaced, 424 families from 18 villages moved to the colonies. Additional 104 families in two villages from the aforesaid 60 villages also moved there in 1991. The break-up of 424 families is given in Table 6.4.

Thus, villages of Biriguda and Parajapondi, Khilaput (Koraput) and Kharaput only shifted to the colonies en-block. Response from other villages was poor.

Table 6.3
Distribution of Land in Colonies (In Acre)

S.No.	*Colony No.*	*Irrigable land.*	*Non-irrigable land.*	*Homestead land.*	*Other communal.*	*Total*
1.	1&2	340.00	222.00	78.00	124.00	764.00
2.	3	232.00	7.00	59.00	155.00	453.00
3.	4	1,147.00	682.00	344.00	647.00	2,820.00
4.	5	475.00	257.00	124.00	291.00	1,147.00
5.	6	275.00	327.00	114.00	383.00	1,099.00
6.	7	479.00	1501.00	293.00	869.00	3142.00
	Total	2948.00	2996.00	1012.00	2469.00	9425.00

Table 6.4
***Colony*-wise Resettlement of Displaced Families**

Phase	*Name of submerged village from which shifted.*	*Total No. of families displaced.*	*No. of families shifted to colony.*	*Colony number where resettled.*
1st	Semela	14	5	5 &6
1st	Kechela	171	18	5
1st	Koranga	322	1	7
2nd	Paidaput	65	21	5
2nd	Landiguda	78	5	5
2nd	Balda (Routput)	113	14	5
2nd	Gopalput	141	11	7
2nd	Tiadiput	95	7	5
3rd	Biriguda	28	28	4&7
3rd	Doraput	63	41	7
3rd	Haldiput	29	2	5
3rd	Kuturiput	24	3	5
3rd	Parajapondi	227	227	4
4th	Khilaput (Koraput)	21	21	6
4th	Khaparaput	3	3	6
4th	Pakhanput	7	1	7
4th	Tiadiput	26	15	7
4th	Khilaput	1	1	6
	Total	1428	424	

Reasons of Reluctance

W. Partridge, Resettlement Specialist of the World Bank in his "suggested outline and recommendations for an action plan for resettlement and rehabilitation in the Upper Indravati Project" writes as follows:-

> "Global experience with the social process of population resettlement indicates that between 10 and 20 per cent of a displaced community will choose to move on their own rather than a Government scheme, no matter how good the scheme is, and these are always the wealthier, better educated, leading families. Their main interest is increasing the amount of money given to them by Government, hence their leadership talents are devoted to protest and resistance. All resettlement options presented will be rejected, and the majority of the villagers will have little voice in such a decision. On the other hand, global experience also demonstrates that when a viable resettlement plan is offered which has the promise of improving conditions of the poor and very poor, then this majority will reject the leadership of the dominant families and cooperate with Government."

In Upper Kolab Project, out of 3,171 families declared as oustees, 424 families, moved first to the colonies and 104 additional families moved later making the total 528. The ramaining 2,643 families which constitute 83.34 per cent of the total displacement have chosen to move on their own.

Partridge is very optimistic in his statement that when a viable resettlement plan is offered which has the promise of improving conditions of the poor, the oustees cooperate with Government and move to the colonies. In Orissa, in case of Rengali Multipurpose Project, Partridge would find himself wrong in his assessments. The rehabilitation measures provided there cannot be termed to be very much viable even after lapse of several years. In spite of it about 70 per cent of the displaced families of Rengali Project have been shifted to 61 colonies, 95 clusters and 652 other locations in existing revenue villages by way of allotment of land. The committee on Estimates of Orissa Legislative Assembly, 1986-87, in course of visit to the resettlement colonies detected innumerable lapses there and in their 37th report made the following observations:-

> "Employment avenues should be created for the people of the area by implementing anti-poverty programmes."
>
> XX XX XX
>
> "the project authority should come forward for constructing model villages as per the norms fixed."
>
> XX XX XX
>
> "though the families of the resettlement villages were assured of irrigation facilities after allotment of lands coming under the ayacut area of Rengali Project have not yet got the benefit of Irrigation."
>
> XX XX XX
>
> "each such resettlement village should be connected with a good road preferably a fair weather road for the transport and communication."
>
> XX XX XX
>
> " Drinking water facilities should be provided excavating tanks exclusively for drinking water purpose."
>
> XX XX XX
>
> "to make special arrangements for medical facilities in some form or other to be available in all the resettlement villages."

There were many debates on the rehabilitation aspects of Rengali Project in the State Assembly; so much so that the House constituted a Special Committee on rehabilitation under the Chairmanship of Hon'ble Speaker. The Committee was very critical about the rehabilitation arrangements provided by the project authorities of Rengali. May be that the oustees had no alternative other than shifting to government locations.

The situations prevailing in Upper Kolab Project are different. They are more in the line of observations made by Partridge that for lack of viability, in the resettlement plan, the response from the oustees to shift to project sponsored colonies was poor. Hence, a majority of the oustees irrespective of the fact that they were wealthier/educated or not, moved away from the settlement colonies on their own. Let us analyse the reasons of reluctance of the majority of the oustees to move to settlement colonies set up by the U.K. Project.

(a) Loss of 'BHITAMATI'

There is always a tendency in human being to react when The suffers. It is more so in case of poor, backward and less educated people. One thing common which unites all the displaced persons in this regard irrespective of the fact whether one is a landlord or landless is the apprehension of submergence of their *'BHITAMATI'* (Birth Place), however, small and poor in value it may be. The prospects of the reservoir project forced the oustee to leave his 'BHITAMATI'. There was no alternative than choosing to shift to a place nearer to the 'BHITAMATI' instead of moving to settlement colonies set up at far off places.

(b) Everybody Equal

Like 'Death the great leveller', the submergence levels the landlord and landless and make all of them equal in the new colony set up for their settlement. Each displaced family is given equal extent of land/rehabilitation cash grant (excepting the compensation money towards acquisition of land) irrespective of pre-submergence financial and social status. The erstwhile big land owners, who had virtually retained the poorer oustees as their bonded labourers shunned to move to the colonies. It is indeed a great social change. The old master-labour relationship is now non-existent in the colonies.

(c) Tree-Culture

The economy of most of the oustees of U.K. Project was dependent on tree-culture. Tree was everything to them. With gradual denudation of forest growth, many of them pushed themselves further into deep and higher areas. They were apprehensive that the forest environment lost to them because of submergence, would no longer be available to them in the settlement colonies.

(d) 'Podu' Cultivation

Most of the tribals believed in 'Podu' (slash and burn) cultivation. They were, therefore, in the habit of shifting to nearby forest areas in search of new cultivation. The resettlement colonies were set up at a distant place, more than 50 Km away from the submergence area. Hence these tribals faced a great psychological insecurity to move out to long distances.

(e) Bad Economic Condition

Economic condition of most of the oustees was very poor. They could not feel themselves confident to be able to make out a better living in the colonies where lands were not that good and needed considerable investment which they could not afford. This aspect prompted most of them to make their own arrangement as they liked.

(f) Difference in Culture

The oustees mainly belong either to 'Paroja' or 'Godaba' community whereas the host population in and around the colonies mostly blong to 'Bhatoda' and 'Amantya' community. While the former maintain their culture in isolation, the latter being mainly agricultural in profession are accepted as being within the Hindu fold. Some of them even wear sacred thread. The oustees, therefore, felt diffident to merge into the new social order.

'Parojas' still dominate in the resettlement colonies. But the 'Gadaba' oustees have run away to hill tops as they did not like to be rehabilitated in the new and unfamiliar of settlement colonies. Some of them have also adjusted with their counter parts living in other unaffected villages in the catchment area.

(g) Risks of Environment Change

The oustees who were the dwellers of hills and forests felt utmost despondent to lead ceremonial life in new colonies down in comparatively lower altitude. They could imagine that the transfer of experience from forest environment to plains environment may be difficult and full of risks. They apprehend that they will bid goodbye to a simple life in close communion with nature and will have to lead a life which will be more detached from nature.

(h) Inadequate Civic Amenities

Last but not the least is the inadequate civic amenities provided in the colonies. Most notable of them is the lack of drinking water facilities. It is hardly surprising that the oustees rejected the colonies where land was infertile, potable water was inadequate and living conditions were different. In the words of Mahatma Gandhi-

> "The dwellers are summarily ejected and are expected to shift for themselves. They are paid a paltry vacating expense which carries them nowhere. Their occupation is gone. They have to build their cottages and search for their livelihood.

(i) Cash in Lieu of Land

But the most important reason which prompted the displaced persons not to move to the project sponsored colonies is the decision of government to make payment of cash grant in lieu of land. It has been said earlier that rehabilitation measures adopted in case of oustees of Rengali Multi-purpose Project have been adopted in subsequent projects. In Government of Orissa, Irrigation and Power Department Letter No. 7213-224/83 dt. the 11th February 1983, it was communicated to the Deputy Commissioner, Rehabilitation, Rengali Dam Project that;

"in view of the restrictions on use of forest land, it has been decided that wherever land is not available for settlement of displaced families, rehabilitation grant shall be paid to such families against land to be allotted as per existing policy, at the rate of Rs. 2160 per acre of unirrigated land. In all cases where rehabilitation grant is paid in full, grant for the house site shall also be paid at the above rate so as to enable the displaced family to settle at the place of his choice."

"Wherever land is available for partial allotment to displaced families, every landless family of the submersible area shall be allotted 1.25 acres of irrigated land or 2.5 acres of dry land. For displaced families losing land upto 6 acres in the submersible area, the allotment shall be 75 per cent land lost subject to a minimum of 1.25 acres of irrigated land/2.50 acres of unirrigated land and maximum of 2.5 acres of irrigated land/4.50 acres of unirrigated land. In allotment of land priority shall be given to displaced landless families. The land allotted shall as far aspossible be rounded off to the nearest 25 cents or its multiple. Where partial land allotment is made, the rehabilitation grant shall be payable for the extent of land not allotted against the prescribed area."

"In all cases of payment of any rehabilitation grant, the Revenue Divisional Commissioner, Northern Division would take steps to persuade the displaced families to deposit at least 75 per cent of the grant under the small savings deposit scheme."

At the relevant point of time, irrigable land upto 3 acres or unirrigable land upto 6 acres and homestead land upto 0.50 acre was being allotted. Those oustees willing to receive rehabilitation grant in lieu of land were entitled to receive Rs. 14,040 in total at the rate of Rs. 2160 per acre of unirrigable/homestead land. This land ceiling was later amended to make 5 acres of unirrigable land or 2.50 acres of irrigable land in the maximum. The monetary celing of Rs. 14,040 was also enhanced to Rs. 20,075 at the rate of Rs. 3650 per acre. It was further amended in August 1990 to Rs. 8000 per acre of unirrigated land and Rs. 16000 for irrigated land subject to certain conditions. However, these amendments have not been adopted in Upper Kolab Project as per decision of Government.

The Commissioner-cum-Secretary, Irrigation and Power Department, Government of Orissa in his D.O.letter No. 16043 dt. 1.5.1979 to the Revenue Divisional Commissioner, Southern Division intimated the Government decision to implement the rehabilitation policy for Rengali Multi-purpose Project, for Upper Kolab Multi-purpose Project too.

The non-availability of Government land especially forest land for non-forest use was the consideration for payment of rehabilitation cash grant. The enforcement of Forest (Conservation) Act, 1980 put a lot of checks and balances in the matter of availability of forest land for non-forest use. It stipulates that for getting clearance of the Government of India for execution of a project, provisions should be made to compensate for the loss of forest land. Where nonforest lands are available, compensatory afforestation be raised over equivalent area of non-forest land; and where lands are not available, compensatory plantation be raised over degraded forest twice in extent to the area being diverted. It further speaks in clear terms that Government of India will not allow diversion of forest land for rehabilitation of people.

Upper Kolab Project being a project which was administratively approved before 1980, the Forest (Conservation) Act, 1980 is not applicable to it. Besides adequate extent of land was already available in Boriguma, Kotpad and Sasahandi area to develop resettlement colonies. But the decision of Government to pay rehabilitation cash grant in lieu of land changed the position. On receipt of specific proposals from the Revenue Divisional Commissioner, Southern Division, Berhampur, Government in Irrigation and Power Department in letter No. 9769 dt. 14.3.85 conveyed that displaced persons not willing to shift to colonies and opting to receive rehabilitation grant cash in lieu of land must give a written understanding to that effect and this must get the approval of the Rehabilitation Advisory Committee. The matter was discussed in the

8th meeting dt. 15.5.85 of the said Committee. But as the minutes of the meeting reveal, due to lack of time to get the approval of the committee, cash payment had already been resorted to in anticipation of approval and as a matter of routine only, the Committee ratified such action in the said meeting.

In such uneasy situation, cash payment in lieu of land was made. Firstly the displaced persons showed their extra enthusiasm to be on their own. They were carried away with the prospects of possessing heavy cash in preference to permanent gain of land. Secondly the Government machinery in charge of resettlement and rehabilitation considered it to be very safe arrangement to make cash payment and be relieved of the worries of rehabilitation arrangement to be made by them. One would call it hypocracy—to call the rehabilitation policy very liberal because of the land base, and make cash payment in lieu of land due to its non-availability.

Ultimately out of total 3171 families displaced, 528 families only shifted to colonies and 2643 families moved on their own to place of their choice.

Economic Structure

(a) Land Holding

The Table-6.5 gives the details of land holding of 424 families resettled in different colonies.

The Table 6.5 the details of land holding of 424 families resettled in different colonies.

These figures do not include 0.50 acres of homestead land allotted to each and every resettler family. The word "irrigable" here refers to the land that will be irrigated when the canal systems of Upper Kolab Project are fully developed. One acre of irrigable land is computed to be equal to two acres of non-irrigable land. Of the 424 families resettled in four colonies, 314 families were given irrigable land only of different extent, 74 families were given non-irrigable land only of different extent and 36 families only were given part irrigable and part non-irrigable land.

Table 6.5
Agricultural Land Holding (in acres) in Colonies

Colony No.	*Irrigable land*	*Non-irrigable land.*	*No. of families.*
	2.00	-	226
4	2.00	1.00	3
	1.50	2.00	7
	1.50	1.00	1
	1.00	3.00	2
	1.00	2.00	2
			241
	2.00	-	23
	1.50	1.00	1
	1.00	1.00	1
5.	0.75	1.50	1
	0.70	1.30	1
	0.50	1.75	1
	0.00	4.00	6
	0.00	3.00	17
	0.00	2.00	2
	0.00	1.50	1
	0.00	1.00	2
			56
	3.00	-	3
	2.75	-	3
6.	2.62	-	9
	2.50	-	9
	1.50	-	6
	-	3.00	1
			31
	2.00	-	17
	2.00	1.00	1
	1.95	0.10	1
	1.90	0.20	1
7.	1.62	-	5
	1.50	3.00	2
	1.50	-	2
	1.25	-	1
	1.17	1.66	1
	1.00	3.00	1
	1.00	2.00	1
	1.00	1.25	6
	1.00	1.00	1
	1.00	-	4
	1.12	-	1
	0.75	-	2

(Contd.)

Table 6.5 (Contd.)

Colony No.	*Irrigable land*	*Non-irrigable land.*	*No. of families.*
	0.50	-	1
	0.40	-	1
	-	5.00	7
	-	4.00	23
	-	3.00	1
	-	2.50	1
	0.50	5.00	1
	-	1.00	12
	0.37	-	1
	-	0.75	1
			96

(b) Part Land and Part in Cash

The Table 6.6 indicates the break up of families in colonies who received part cash payment of rehabilitation grant money and part allotment of land. This is in addition to the compensation money towards acquisition of their land and property.

Table 6.6
Rehabilitatoon Grant Money Received by Resettlers

Amount per family (in Rs.)	*Number of families*			
	Colony No.4	*Colony No.5*	*Colony No.6*	*Colony No7*
1000 to 2000	-	-	12	-
2160	13	-	9	9
4320	228	30	-	43
4500 to 6000	-	-	-	12
6480	-	19	7	5
7020	-	1	-	-
7100 to 7500	-	1	-	-
7501 to 8200	-	-	-	2
8640	-	2	-	4
9720	-	1	-	2
10,800		2	-	13
10,801 to 11500	- -	-	-	3
Total	241	56	28	93

The tribals are not keen to save for the future. They have got very little urge to acquitre property; and they are more accustomed to do only the minimum of work that will bring them food sufficient for the day only. The idea of working for gain is quite foreign to them. Drinking is an integral part of their life. Keeping all these factors in view, the State Government posted an officer from the Harijan and Tribal Welfare Department as Special Officer, to motivate the displaced persons not to while away the money received by them. The Rehabilitation Advisory Committee in its 8th meeting dt. 15.5.1985 also took a very conscious decision that the oustees will not be paid the entire lot of rehabilitation grant in cash. Instead the following arrangements will be made.

(i)	Deposit in six yearly National Savings Certificate	Rs.6,000.00
(ii)	Deposit in Five Yearly Fixed Deposits	Rs.6,000.00
(iii)	Deposit in Savings Account	Rs.1,500.00
(iv)	Cash payment	Rs.540.00
		Rs.14,040.00

Since compensation money towards acquisition of land and structure etc. was also available with an oustee, there was absolutely no difficulty in implementing the above scheme.

The project authorities have not kept any record about 2643 families who received full cash payment and moved as they liked. But a test check conducted with three families resettled in colonies revealed that Leli Hantal, Sunia Guntha and Tulasi Bairagi had received cash payment of Rs.4320, Rs.4320 and Rs.6480 respectively at the time of shifting in 1987. In November, 1989, they had got Rs.27, Rs.24 and Rs.24 only respectively in their pass books. They had obviously not invested any money on any scheme which would fetch more dividends; because on query, all of them simply smiled and avoided to answer. I.U.B.Reddy, a faculty member of the Tata Institute of Social Sciences, Bomaby had conducted a survey in 1985 in respect of 121 S.C. and S.T. oustee families and 250 caste Hindu oustee families to have an idea about the manner of expenditure of compensation money received by them. The result of his study is reflected in the Table 6.7

Table 6.7
Manner of Expenditure of Compensation Money by Oustees

		S.C.	*S.T.*	*Caste*	*Hindus*
S.No.	*Items of expenditure*	*Amount spent (in Rs.)*	*Percentage*	*Amount spent (in Rs.)*	*Per-cent-age-*
1.	Purchase of land	3,737.00	2.00	-	2.4
2.	Payment of loan	42,599.00	22.80	-	10.5
3.	Purchase of food items.	77,913.00	41.70	-	19.7
4.	Purchase of household items	21,113.00	11.30	-	15.5
5.	Deposit in banks.	19,805.00	10.60	-	22.9
6.	Others	21,673.00	11.60	-	29.0
	Total	1,86,840.00	100	4,79,240.00	100

(*Source* - "Costs of river valley projects in India" by I.U.B.Reddy and "Impact of Large Irrigation Projects on the tribals" by I.U.B. Reddy).

It would be seen from the statistics that a very small percentage of compensation money was utilised for acquiring land or deposit in the banks. It is more conspicuous in case of S.C. and S.T. oustees. Being mostly illiterate and not materialistic, they could not utilise the money in a proper way. They spent the money mostly on liquor and purchasing fashionable articles like transistor radios, wrist watches, dresses etc., Reddy observes that "the money lenders exploited the situation to realise their debts and also demanded high interests."

The author made a sample survey in respect of 85 families out of total number of 96 families resettled in colony No.7, to have an idea about their economic status. The format devised for this is in Annexure-VIII. The Table 6.8 will give the details of agricultural implements and livestock position in the colony.

Table 6.8
Agricultural Implements and Livestock Position

	Agricultural implements			*Livestock*		
S.No.	*Item*	*Posses-son be-fore dis-place ment*	*Posses-sion in colony*	*Item*	*Poss-ession before displa-cement*	*Posses-sion in colony.*
1	*2*	*3*	*4*	*5*	*6*	*7*
1.	Plough	70	60	Cow	334	151
2.	Bullock cart	-	-	Bullock	165	131
3.	Sickle	89	90	Buffaloes	117	63
4.	Axes	95	94	Goats	301	42
5.	Spade	79	89	Sheep	81	22
				Poultry	345	96
	Total	333	333		1343	505

It would be seen that although there is very little variation in respect of number of sickles, axes and spades, there has been a reduction in the number of ploughs. This is primarily due to the fact that the bullocks which are used as plough pullers have reduced in number. Reduction in the number of cow, buffaloes, goats, sheep and poultry is also very conspicuous. The primary reason for this is lack of grazing ground nearby and change in environment from hills to plains. The resettlers instead of replenishing the loss, became more concerned in acquiring other consumables, fashionable articles, clearing of debts and investing on house construction etc. If this is the state of affair in respect of colony No.7 which is recorded to be the best colony in all respects in U.K. Project, less to speak about other colonies will be better.

Thus, the purpose for which the deposit scheme was introduced, did not bear the desired fruits. The project authorities appeared to have remained as passive onlookers, pleading helplessness. Required amount of motivation seemed to be lacking.

(c) Indiscriminate Allotment of Land

The agriculturists in our society are land owners, of whatever size their holdings may be; either by way of inheritance or by way of purchase. The resettlers in case of Upper Kolab Project are land owners, who do not belong to either category. In their case land has been thrust on them. Government allotted land to each individual family whether he was landless or not before displacement. It was also not considered whether the landless oustees were in a position or not to cultivate the land on their own. Table 6.9 is a representative sample of some of the landless oustee families who have been allotted land in the resettlement colonies and the rehabilitation grant received by them in cash.

All these oustees did not get much of compensation money towards acquisition of their property; because they did not have got any agricultural land recorded in their favour. For the homestead land which was "Parambok" land, no compensation money was given to them excepting few chips towards the structures built on the "parambok" land. It is good that government made the landless, the land owners. But it was desirable to evaluate as to if the rehabilitation grant received by the oustees could pull them on or not to acquire agricultural implements and other accessories so as to carry on cultivation in the new set up. Least to speak about the transition period from shifting to raising of crops in the settlement colonies.

(d) No Uniformity in Land Allotment

Further, no uniformity was maintained in allotment of land to the resettlers. Table 6.5 already indicates the size of holding of all the 424 families in four colonies. No thought has been given as to whether land allotted to a family will be economically viable for him or not. By no standard of imagination one acre of unirrigated land will support any family without any subsidiary occupation. Two families in colony No.5 and 12 families in colony No.7 were allotted one acre each of unirrigable land. One family is also in possession of less than one acre (0.75) of unirrigated land and three fami-

Table 6.9
Landless Oustee Families who Received Land/Cash

S.No.	*Resettler who was landless before displacement.*	*Extent of land allotted in settlement colony (in Acre)*		*Colony numbers*	*Rehabilitation grant received (in Rs.)*
		Irrigable	*Non-irrigable.*		
1.	Smt. Amina Bibi	2.00	—	4	4,320.00
2.	Sri Dullabh Benia	”	—	”	”
3.	Sri Durjan Benia	”	—	”	”
4.	Sri Tilak Machadi	”	—	”	”
5.	Sri Dambaru Jani	”	—	”	”
6.	Sri Govind Khara	”	—	”	”
7.	Sri Gobardhan Deulia	”	—	”	”
8.	Sri Balaram Khemudu	—	4.00	5	4,320.00
9.	Sri Bhimasen Panda	—	”	”	2,160.00
10.	Smt. Ramadei Golari	—	3.00	”	6,480.00
11.	Sri Laikhan Hantal	2.00	—	”	4,320.00
12.	Sri Madhu Hantal	”	—	”	”
13.	Sri Sania Guntha	”	—	”	”
14.	Sri Keshab Majhi	3.00	—	6	nil
15.	Sri Dayanidhi Nayak	”	—	”	nil
16.	Sri Bhagaban Nayak	”	—	”	nil
17.	Sri Chitrasen Patra	2.75	—	”	4,320.00
18.	Sri Raghuram Patra	2.50	—	”	”
19.	Sri Ramakrushna Patra	”	—	”	”
20.	Sri Sanu Burudi	”	—	”	”
21.	Sri Arjun Majhi	”	—	”	”
22.	Sri Sukra Gulel	2.00	—	7	4,320.00
23.	Sri Arjun Khara	2.00	—	7	4,320.00
24.	Sri Naran Talab	”	—	”	”
25.	Sri Debraj Maharana	”	—	”	”
26.	Sri Damburu Talab	1.17	1.66	”	”
27.	Sri Balram Gabang	—	4.00	”	”
28.	Sri Madhu Talab	—	”	”	”
29.	Sri Gourang Gabang	—	”	”	”
30.	Sri Mukund Bhoi	—	”	”	”
31.	Sri Dhanurjay Khila	—	”	”	”
32.	Sri Kunda Katur	—	”	”	”
33.	Sri Sukadeb Dora	—	1.00	”	”

lies have got less than one acre each of irrigable land in colony No.7 The two families in colony No.5 have got six members each. As regards the twelve families in colony No.7, their family size varies from three members to five memebrs. It is everybody's guess to gauge the economic predicament of these families.

(e) Quality of Land

In a country like India, which is primarily agrarian, land still represents a means of production, especially when it is lost due to submergence. The Rehabilitation policy of the State Government, therefore, aims at making the resettlers dependent on land. Excepting casual labour, the opportunity for other means of livelihood is also limited. Hence land made available should be of good quality. It is good that government did not forget that the oustees who sacrificed a lot for the construction of the project, should get the benefits of irrigation. As a matter of fact about 50 per cent of the agricultural land allotted in four colonies are irrigable. It has been scheduled to complete the canal system and bring water to this area by 1991. But there is many a slip between the schedule made and the result achieved. The author had been closely associated with the project activities during his tenure in the Irrigation Department of State Government. A considerable patch of reserve forest land intervenes the Jeypore Main Canal beyond 49 Km distance. Unless and until clearance is obtained from the appropriate quarter to utilise this patch of forest land, canal water cannot pierce into the colony areas within the time scheduled. Earnest effort has to be made in this regard. The resettlers have started feeling restless for the want of canal water. Any further delay will invite problems and nothing but problems.

Government spent a lot of money on reclaiming the land identified for the resettlers. But reclamation has not been done properly. Tree stumps were not uprooted and for which at several places, tree growth has again appeared.

Thirdly, the land selected is dominated by laterite soils of low fertility, yield is, therefore, not encouraging. The majority of the resettlers who were surviving by slash and burn cultivation are not very confident that the land allotted to them will enable them to carry on.

It is unfortunate that there was not activist organisation here which could champion the cause of the oustees. It is unfortunate because, not that all the activists' organisation are anti-government; but because most of them also place the genuine difficulties of the people, for timely redressal. On the other hand, the peoples' representatives including the Rehabilitation Advisory Committee advocated for payment of rehabilitation cash money in lieu of land so that the oustees would make rehabilitation arrangements as they liked. Hence very small percentage of the oustees moved to the colonies and the rest apprehending that the allotted land will not help them in leading a viable farming profession, moved on their own; might be deep into the forests. The irony is that the government machinery is not successful in stopping the oustees to do so. There has also been no attempt to trace their present status.

(f) Encroachment

In comparison to the arrangements made in the colonies, less number of families moved there. Consequently the settlers got scope to encroach land which was available free for all. On interview, several resettlers made no hesitation in admitting encroachment by themselves too. There seemed to be a point in their argument as under:-

> "One acre of irrigable land is equivalent to two acres of unirrigated land. Instead of allotting irrigated land, Government allotted irrigable land with assurance that irrigation will be available soon. If irrigation has not come so far, there is no wrong in cultivating additional land beyond allotment so that it could be twice in extent of the allotted irrigable land."

It is good that Government have remedied this anomaly in Irrigation Department Resolution No.31888 dt. 21.8.90 to the effect that land whether irrigated or unirrigated at the time of allotment, will be taken into consideration, without taking into account whether there is future scope of irrigation facilities or not.

The Table 6.10 gives the details of land particulars in four colonies.

Table 6.10
Land Particulars in Colonies.

Colony No.	*No.of families resettled.*	*Land allotted (in acre)*			
		Homestead.	*Irrigable*	*Non-Irrigable.*	*Total*
4	241	120.50	479.00	39.00	638.50
5	56	28.00	51.75	75.50	155.25
6	31	15.50	72.33	3.00	90.83
7	96	48.00	69.40	173.91	291.31
Total	424	212.00	672.48	291.41	1,175.89

The Rehabilitation Advisory Committee in 10th meeting dt. 25.4.88, recommended that 104 number of additional families of two villages namely Khilaput and Khapraput and who would be affected by the reservoir water should be rehabilitated in the resettlement colonies. This was accepted by Government in Irrigation and Power Department. Accordingly the project authorities have worked out the details to settle these additional 104 families in colony No.7 only by allotting 50 decimals of homestead land, 2.00 acres of irrigable land and 2.00 acres of unirrigable land to each family. In that event, for these 104 families additional 52 acres of homestead land and 208 acres each of irrigable and unirrigable land will be required. This would mean that statistically land in different colonies would be available in surplus as in Table 6.11.

This land is surplus to the requirement, as less number of families than estimated shifted to the colonies. For effective utilisation of this surplus land, Government in Irrigation and Power Department have taken decision to shift some of the displaced families of Upper Indravati Project. But the following points need consideration:-

Table 6.11
Surplus Land Available in Colonies (in acre)

Colony No.	*Irrigable*	*Non-irrigable*	*Homestead*	*Others*	*Total*
1 & 2	243.65	69.24	64.86	223.57	601.32
3	253.87	34.58	48.50	115.32	452.27
4	668.46	600.12	103.32	-	1,371.90
5	223.25	434.37	47.00	-	704.62
6	213.65	289.94	83.00	-	586.59
7	186.38	975.52	92.50	-	1,254.40
Total	1,789.26	2,403.77	439.18	338.89	4,971.10

(i) Land shown in paper to be surplus may not be found physically surplus. Firstly, some of the host population grabbed some land in their anxiety to get irrigation facility when available. Secondly several resettlers themselves have encroached land as the same was available close at hand.

(ii) Upper Indravati Project authorities will not take possession of any land which is not free from encumbrances. A very tough time is there ahead, for the Upper Kolab Project authorities to resolve the matter. Confrontation between the host population and resettlers and between two groups of people and oustees of Upper Indravati Project cannot be ruled out. It has to be tackled very intelligently and with all nerves.

(g) Distribution of Patta

"Patta" vests right and title of an individual over the land. It gives power and authority for sale, transfer and mortgage, etc. The resettlers have been given patta so far for the homestead land only. There is no satisfactory reason for non-distribution of patta in respect of ágricultural land. In the result, the settlers are not in a position to avail of loan facilities, either short term or long term. Nobody has stood for their cause. As the author foresees, there will be further delay in distribution of patta for the reason that forest land taken to the fold of the colonies has to be dereserved after following the laid down procedure.

(h) Cropping Pattern

The staple food of the oustees was rice and ragi. It also continues to be so of the resettlers. Irrigation is long way to come. Hence rainfed agriculture is the only source to bank upon. The crops grown are short duration paddy, ragi, millet, maize and niger. Improved method of cultivation or capital intensive cultivation are not known in their farming system. Agriculture operation starts from 15th June anticipating rains to follow. As a matter of fact, the resettlers are seeing the staple crop for the second time only after their shifting during 1987. Laterite soil, lack of irrigation facility and financial inability of the resettlers to take to improved method of agriculture result in poor yield. Sania Guntha, a resettler of colony No.5 was harvesting his paddy crop on 15.11.89. He was visibly happy to harvest paddy crop from his own land, because he was landless before evacuation. But the statistics given by him were indeed shocking. "The Economic Survey," 1988-89, published by the State Government in Planning and Coordination Department gives the statistics that in a drought year like 1987-88, the yield rate of paddy in the State was 8.56 quintals per hectare. 1988-89 was not a bad year from agriculture point of view. But in the words of Sania Guntha, he got about two bags *i.e.*, about one and half a quintal of paddy from one acre of land. This by any means is a disheartening state of affair. It is unfortunate that the State Agriculture Department have paid no attention to improve the position. Consequently the resettlers have resolved so far to their fate.

Keeping the soil conditions and lack of irrigation facility in view, the Agriculture Department should have prepared a suitable cropping pattern. The available agriculture extension service should have been duly activated in this regard. Pilot programmes for horticulture, fodder and cashew plantation could have been organised at suitable patches. This would have also helped the resettlers in educating them adequately to take up alternative occupational avenues.

Equal attention has also to be paid to the oustees who have resettled themselves on their own in the upper catchment area. They are required to be detracted from practising shifting cultivation and intead, provided with gainful employment and alternative agriculture land. Extensively horticultural development is feasible because of the climatic and water potentiality of the area.

Occupational Structure

(a) Agriculture is the Principal Occupation

Most of the resettlers earned their living before being displaced from agriculture whether directly or indirectly. In the settlement colonies, every one of them including the previously landless has been allotted some extent of agricultural land either irrigable or non-irrigable, or both. Hence all the resettlers are now dagriculturists on record.

There is no record to know about the annual income of the oustees before their displacement. With two cultivation seasons only in colonies, the resettlers have not been able so far to have a firm foothold on land-based economy. Their calculations are erratic and without any arithmatics. The trauma of displacement has preoccupied their mental state so much so that they are very much adverse to any question regarding their well being. However, realistic or unrealistic it may be, the information collected from them gives the picture as in Table 6.12.

The Table indicates that there has been increases in the income group of Rs. 4000 and in the group above Rs. 10,000; and fall in the other two groups. With the passage of time and availability of irrigation water as well as motivation for improved agricultural know how, things will definitely improve. They have been given lands, but not the resources. With introduction of various of on-farm development packages, things are apt to improve. This is very necessary to be introduced on priority.

It is indeed significant that the shift to the colonies, has brought in a change in the occupational pattern. Allotment of

Table 6.12
Annual Income from all Sources (Figures in Families)

Income prior to displacement					*Present income*			
Upto Rs.4,000	*From Rs.4,000 to Rs.5,000*	*From Rs.5,000/- to Rs.10,000*	*Above Rs.10,000*	*Co-lony No.*	*Upto Rs. 4,000*	*From Rs. 4,000 to 5,000*	*From Rs. 5,000 10,000*	*Above Rs. 10,000*
181	39	14	7	4	228	4	2	7
20	15	16	5	5	38	11	2	5
17	2	9	3	6	23	3	3	2
78	6	12	-	7	87	-	-	9
Total 296	62	51	15		376	18	7	23

land to each and every resettler family has made all of them to adopt cultivation as their principal source of income. Some of the families have not only cultivated all the lands allotted to them, but have also encroached upon Government land. The trend, however, prejudicial to Government interest may be, is indicative of a change for a living, better than what it was before submergence.

(b) Increase in Homestead Land Holding

Another welcome change for better is the uniformity in allotment of homestead land at the rate of 0.50 acre per family. Before submergence, the families on an average had not even got ten cents of homestead land. Thus, additional homestead land at their disposal would enable them to make provision for meeting the requirements of their growing families in future.

(c) Employment

In 1974, Government in Irrigation and Power Department took a decision that 50 per cent of the vacancies in unskilled posts in the project will be kept reserved for the people of submersible area. But this decision has not been implemented faithfully in Upper Kolab Project. Only 25 displaced persons have been employed in the project in Class IV posts and the number of N.M.R. employees from among the oustees is 23 only. These N.M.R. employees do not have got any permanent footing; as they are likely to be retrenched in course of reduction of construction activities in the project. Eight number of persons residing in colony No.4 are service holders in organisations other than U.K. Project. One person in Colony No.5 is a petty contractor.

No significant occupational skill has flourished in any of the resettlement colony because of the fact that either the resettlers are not enterprising enough or the Government machinery has not taken adequate interest in this regard. During the plan developments especially in the Seventh Plan, special emphasis was given to I.R.D.P. (Integrated Rural Development Programme), E.R.R.P. (Economic Rehabilitation of

Rural Poor), N.R.E.P (National Rural Employment Programme) and TRYSEM (Training of Rural Youth for Self Employment). It is a pity that none of these programmes has been tried in the colonies. The greatest bottleneck in this regard is that the resettlers' households have not been enlisted so far in the normal block development activities. So much so that the resettlers have not got the opportunity to exercise their franchise during the last two elections for the Parliament and Assembly.

(d) Cry for Wage Earning

Land allotted to the resettlers is not fully developed. Irrigation will take time to come. Development activities from the Government side especially to provide employment are non-existent. As with most welfare oriented schemes, the resettlers in U.K. Project have tended to develop dependency on Government and of complaining attitude. This is mostly because of the fact that there is lack of employment opportunities. Construction of canal system of the U.K. Project, in and nearby the colonies was in good speed. But as complained by several resettlers of colony No. 5, they were not allowed by the original villagers to work for any contractor which will jeopardise their own employment prospects. On the other hand it should not be forgotten that the resettlers are basically timid and shy. They have not been able to adjust themselves to the purely agriculture based social environment at the new place. They lacked the requisite speed and skill in cultivation, least to speak about earth work for the excavation of canals. Hence the contractors hesistated to exploit the potential labour of the resettlers. It is unfortunate that the project authorities have not paid any attention to this aspect. Displacement of population has been caused due to construction of the project. Gainful rehabilitation of these people is the responsibility of project authorities. Contractors employed by the project should have been impressed upon to move in the right direction to employ the resettlers.

This indicates that the resettlers seem still alien to the host community and no conscious effort has been made either by the project authorities or by the District administration to foster a community spirit among both the hosts and the resettlers.

(e) Hard Transition Period

An important aspect which has been lost imagination of the Government is the need for arrangements during the transition period from the date of shifting of the oustees till first harvest of the crop by them in the new place. The State policy did not provide then for payment of any exgratia/subsistence money for this period. (As per irrigation Department Resolution No. 31888 dt. 21.8.90 maintenance allowance at the rate of Rs. 500 per family per month for one year will be given to oustees residing in project sponsored colonies). In the alternative, some thought should have been given to provide the resettlers with an income generating scheme to fall back upon. Employment in installation of infrastrcture, preparation of agricultural land, growing of woodlots, etc, could be the possible alternatives. Vocational training programmes with provision for payment of allowances could also have been organised. This would have helped some of the resettlers, in earning some money as well as in making them eligible to work in various Government as well as non-government organisations.

(f) Easy Money to Spend

Experience in other water resources projects in Orissa indicates that there is a wide time gap between payment of compensation money and shifting of the oustees. In Upper Kolab Project, there was not much of delay in shifting, because there was no element of compulsion unlike projects previous to Upper Kolab Project, to shift to the settlement colonies. Here the oustees knew it all right that there was no compulsion to move to the colonies. So much so that the Rehabilitation Advisory Committee in the 8th meeting dt. 15.5.1985 accorded *expost facto* approval to the payment of rehabilitation cash grant in lieu of land. Even much prior to it the Committee in the 7th meeting dt. 20.12.1983 decided that "all sorts of cash grants except cost of first ploughing should be disbursed to the D.Ps prior to their movement from the Villages." Thus, all the oustees including those who moved to the settlement colonies were on their own to spend the money freely. The Resettlement and Rehabilitation staff were also not very much keen in their duty. There was, of course, positive lack of response from the resettlers; which is obvious. But a sense of missionary zeal in the officials could have improved the situation. In the alternative, involvement of voluntary social organisations would have gone a long way in guiding the resettlers in making fruitful investment of the compensation as well as rehabilitation money. Consequently, these lapses gave a free hand to the resettlers to spend the money prodigiously, especially in marriages and other social ceremonies which had remained unsettled and were being deferred earlier due to want of adequate funds. It so happened that several settlers married for the second time while first wife was alive. Informations collected indicated that 2,3 and 9 resettlers in colony No. 5,6 and 7 respectively had got more than are wife; and they all married again after shifting from the submerged area.

Population Structure

It has been discussed above that the oustees constituted 45 per cent S.T., 14 per cent S.C. and 41 per cent general. The study revealed that about 14.5 only of the displaced S.T. population shifted to colonies. They were comparatively economically better than other S.T. people who were hardened tribals, not willing to adopt a living in the plains. They, therefore, left in search of fortune in the high attitudes and other forest areas and their whereabouts have not been documented so far. Such category of people tend to be nomads. 'Pana' formed the majority sub-caste in S.C. community. They are known to be very shrewed and intelligent in S.C. community. They are always interested in seeking self- interest. Hardly 3 per cent of them moved to the colonies and the remaining left here and there where better opportunities were available. They had already arranged for it before physically deserting the submerged area. Of the general category of population, 'mali' and 'rana' formed about 90 per cent and they were the comparatively well to do people living on agriculture, trading on fruits, nuts and flower and worshipping the deities in rural areas on consideration mostly in kind. They had their lineal contact in the country side in kind of which they preferred to take advantage. Generally it was ascertained that most of them had also arranged for purchasing land in areas where their kinsmen had lived. The Table 6.13 indicates the number of households, their size and male and female composition in the resettlement colonies.

Table 6.13
Household and Population Structure in Colonies.

Sl. No.	*Colony No.*	*No. of households.*	*Poplation*		*Total population.*	*Average household size.*
			Male	*Female*		
1	4	241	531	513	1044	4.33
2	5	56	113	117	230	4.10
3	6	31	65	59	124	4.00
4	7	96	188	174	362	3.77
Total		424	897	863	1760	4.15

Table 6.14
Size of Families.

S.No.	*No. of persons in the family*	*Number of families*				
		Colony4	*Colony-5*	*Colony-6*	*Colony-7*	*Total*
1	1	6	3	1	9	19
2	2	19	10	8	7	44
3	3	18	11	7	24	60
4	4	70	10	3	23	106
5	5	109	7	2	28	146
6	6	17	9	7	2	35
7	7	2	4	2	1	9
8	8	-	1	1	2	4
9	9	-	1	-	-	1
Total		241	56	31	96	424

Colony No. 4 is a bigger colony having three different segments *i.e.,* 4, 4A and 4B. Hence this colony contains more household size *i.e.,* 4.15 and the male population is slightly higher than the female population. Of course, in colony No.5, female population is having a very slight edge over the male population.

The table 6.14 indicates the actual size of families.

It does not indicate any particular pattern. Single/two members families as well as large families are distributed among all the communities.

Jolt in Joint Family System

Very large families comprising six members or more occur in few cases only. It gives an indication that joint family system is less prevalent. Normally in rural areas and more so among the S.T. and S.C. community, the major sons after getting married become separated and live in separate mess and property. This system is further prompted due to displacement; because to claim eligibility for separate family and get land/rehabilitation grant, major sons putforth their cases. This is how joint family system which has got big tradition and deep roots in India gets a big jolt. In the old village, cooperation, kinship, etc., evolved round joint property. But in the new colony, the hunger for individual property. discards the values of joint family system. The study revealed that very few joint families comprising father, mother, son and daughter-in-law are there. Their number is 11, 3 and 4 only in colony Nos. 4,5 and 7 respectively. Elsewhere the family comprises the male member, his wife minor sons and daughters and dependent mother/father. The family system is patrilineal where the male play the dominant role and the authority is patri-centred.

Ethnic Group-wise Distribution of Population

The table 6.15 gives the ethnic group-wise distribution picture of population.

Thus, the dominant ethnic group is 'Paroja' who belong to Scheduled Tribe. Further they are mostly concentrated in Colony No.4 and all of them came from the submerged village

Table 6.15
Ethnic Group-wise Population (Figures Indicate Number of Families)

S.No.	*Caste*	*Colony-4*	*Colony-5*	*Colony-6*	*Colony-7*	*Total*
1.	Paroja	189	17	-	1	207
2.	Harijan	25	3	-	7	35
3.	Paika	14	12	-	38	64
4.	Muslim	3	-	-	-	3
5.	Rana	3	10	23	32	68
6.	Mali	-	-	8	-	8
7.	Lahara	7	-	-	-	7
8.	Gouda	-	6	-	-	6
9.	Baisnab	-	5	-	-	5
10.	Golari	-	1	-	-	1
11.	Brahmin	-	2	-	-	2
12.	Maharana	-	-	-	1	1
13.	Kshyatriya	-	-	-	17	17
Total		241	56	31	96	424

Parojapondi. 'Rana' are the next dominant group, scattered in all the four colonies. Although agriculture is their main occupation, they belong to the worshipper community in the weaker section, as substitute to Brahmins. 'Paika' belong to the warrior community. They formed the core infantry of the ex-rulers.

Excepting in Côlony No.5, the other colonies are not also multi-ethnic in composition. In any rural set up in India, interdependence especially in rendering caste services is a felt need. But in the resettlement colonies, some basic minimum caste services like washing, shaving, etc., are lacking. Hence the resettlers have to depend for these services on outside people.

The common language of the host population and the resettlers has become Oriya; but there still remains ethnic sensitivities. Intermarriage is still limited. Purchase of brides is still prevalent among the tribals.

Literacy

The informations collected are not true representative, because they speak about the literacy among the heads of the family only. The position relating to the individual members of the family could not be obtained.

Table 6.16
Literacy Position.

S.No.	*Colony No.*	*No. of families.*	*No. of literates*	*Percentage.*
1	4	241	15	6.22
2	5	56	15	26.78
3	6	31	16	51.61
4	7	96	23	23.95
		424	69	

Higher education was not known to them. The maximum education standard known was Class IX in the High School. The break-up of 69 persons so far as educational standard is concerned, is as follows:-

Class I	..	1
Class II	..	14
Class III	..	24
Class IV	..	3
Class V	..	17
Class VI	..	2
Class VII	..	5
Class IX	..	3
		69

Excepting one Kanyashram (Girls' High School run by Harijan and Tribal Welfare Department) which was shifted from the reservoir area to Deoghati, there was no other High School to afford school education of that standard. The school system that existed before submergence and the system prevailing in the settlement colonies, is merely an extension of the national system. The children of the oustees who the other day, were passing their time in the jungle in search of birds to hunt and fruits or fuel to collect, get sore of routine school life. Hence it is very necessary that non-formal education is imparted to them.

Further more and more vocational training facilities, education on awareness towards developed farming system, environment, etc., should also be afforded to the grown-ups.

Health Hazards

The ex-dwellers of hills care little for their health. Any disease befalling them is attributable to the curse of 'Thakurani'. When asked about new diseases appearing in the settlement colony and old diseases had disappeared, the resettlers were not very responsive. Government have not made any provision for health services exclusively for the colonies. Health care facilities are available at Kotpad, which is not far off.

Social Structure

A resettler's life system in a settlement colony has a unique position in the society. Its chief characteristic features are:-

(*a*) It is non-competitive. No other system or organisation challenges it. It enjoys the goodwill of government.

(*b*) It is relatively an isolated social system. It is a structure composed of one community *i.e.,* the "displaced community." People are identified by their evacuation, not by caste, religion and culture.

(*c*) Another special feature of the resettler's life system is that it is a closed or protected system. Until the households are merged into the general administrative system of the district, the resettlers enjoy a protected life. There is absolutely no interference of any outside society. On the other hand the resettler is always apprehensive to merge in the main stream of the social structure.

(*d*) The predominant interest of the resettler is to adjust to the expectations of the host population. Any change in outlook in this regard is likely to create problem for him.

(*e*) Man enjoys his status in the society depending on his income. But the resettlers' society carries very little regard to income; because irrespective of their economic status prior to displacement, they all are placed in equal status in the colony by way of allotment of equal extent of land.

In practical field the resettler gets sore of this system which he calls an "outcaste" system. On the one hand he wants to take advantage of all the special economic benefits that should be available to him; and on the other he wants to be not termed as a "displaced person". Thus, in two minds he spends his time until he is fully integrated in the community to which he is resettled.

Lapses in Infrastrcture Developments

In Table 5.6, details of the infrastructure development works done in different colonies have been given. Colony Nos. 1,2 and 3 have not been inhabited so far; even though the project has spent a lot of money on infrastructure developments. The following is the position with regard to the works done in the remaining four colonies.

(*a*) *Roads*—Internal roads and approach roads have been constructed in all the colonies. But they are not in good condition. Difficulties are encountered in negotiating from and to the colonies as the roads are not made all-weather.

(*b*) *School*—The State rehabilitation policy speaks that one two-roomed primary school shall be provided to villages with less than 50 families. Hence keeping the settler population in view, Colony No.4 should have got two schools. But contrary to this, they have got one school each like that in Colony Nos. 5 and 6. Two High Schools at Sasahandi and Nuagaon have also been set up at the project expenses.

On a working day on 15.11.89 it was seen that the school in colony No.4 was closed. On enquiry it was learnt that the teacher was availing of leave for the last about one month. The kind of leave availed of was only known to himself. The school was under the management of the Block Development Organisation. But the project authorities did not seem to be in proper coordination with the former. Consequently there was difficulty in management of the school.

(c) *Drinking Water Well*—As per the rehabilitation policy, there should be provided one well upto 44 families, two wells for 45 to 70 families and three wells for 70 to 100 families. The wells and tubewells provided in Colony Nos. 5, 6 and 7 are as per the yardstick. But as against the requirement of seven number of wells in Colony No.4, only 5 wells (2 open wells + 3 tubewells) have been provided. Further the most annoying point is that about 50 per cent of the open wells in all the colonies get dried up during summer. In colony No.5, the depth of one well is 50 ft. But on 15.11.89 it had only 2 ft of muddy water. The tubewells bored contain higher percentage of iron, for which the settlers do not use tubewell water for cooking and drinking purpose.

(d) *Tanks*—The rehabilitation policy speaks that there should be one tank irrespective of number of families in a colony. Excepting in colony No.7, tanks have been provided in all the colonies. As a matter of fact, there are as many as three tanks in colony No.4, distributed between 4, 4A and 4B. But here again all the tank sources of water remain dry for several months of the year. The embankments have not been properly compacted. Consequently, the porous material in the embankments does not hold water. Small quantity of water lying in the excavated portion of the tanks dries up even before summer sets in.

Thus, availability of potable water, for the resettlers as well as for the animals domesticated by them is a great problem. No doubt, during high scarcity time, the project authorities have supplied drinking water in water tankers. But no serious thought has been given so far to resolve this acute problem on permanent basis. As it appeared, the wells and tube- wells were located at places without any trial and without taking the settlers into confidence. In course of visit to the colonies, some of the project officials seemed to be whispering that with release of canal water in the canal system, the water table would rise and would be stabilised. Then there would be no problem of water. It is indeed most discernible. It cannot be appreciated that settlers should wait for water till the canals come up.

(e) *Other Civic Amenities*—The rehabilitation policy also speaks that one club house irrespective of number of families and facilities for reconsecration of deities should be provided in each and every settlement colony. But these facilities are totally lacking in the colonies. Upper Kolab Project will supply hydro-electricity to all the nook and corner of the State through the State Power Grid System; but the colonies remain dark. No effort has been made so far to cover the colonies under Rural Electrification Corporation Scheme. There is neither any Public Health Care Céntre nor any Animal Health Care Centre in any of the colonies. The people have to depend for such facilities on Kotpad and Sasahandi which are about five to seven kilometres away. For their daily consumables, the settlers have to visit the weekly market places at Batasana on Sunday, Kotpad on Tuesday, Sasahandi on Friday and Boriguma on Wednesday. Postal and Bank facilities are non-existent in any of the colonies.

(f) *Lack of Maintenance*—School buildings, wells and tanks, internal roads, etc., were constructed in the resettlement colonies about 7 to 8 years ago. By February 1987, out of 2187 families displaced, only 49 families had shifted to colonies. The shifting is yet to be completed. Hence the institutions to provide social services in the colonies have not taken deep roots even now. The assets created have not been taken over by the concerned Administrative Departments. On the other hand the project has not been able to maintain them for want of funds as the project activities have been completed substantially. Consequently most of the assets have deteriorated and they require repair.

(g) *Fuel wood Obsession*—Few days before the visit of the author to the colonies on 15.11.89, a World Bank team visited the area in connection with the resettlement of displaced families of Upper Indravati Project in land that would be spared by the U.K. Project. D.Butcher, Anthropologist was one of the team members. He pointed out about the following two lapses in the economic condition of the resettlers.

(*i*) Supply of fuelwood.

(*ii*) Availability of threshing yard.

With regard to fuel wood, Butcher observed, "oustees from Upper Kolab obtain wood fuel for cooking by digging up the roots of trees and bushes which were removed when the area was reclaimed. This source of fuel is almost exhausted and the 1580 families from Upper Indravati will not be able to find cooking fuel. Bearing in mind that it is the custom of the majority of people among the oustee groups to cremate their dead, the absence of fuelwood means that the Sasahandi area as it is today is untenable as far as they are concerned. A number of steps need to be taken as a matter of urgency to solve the immediate needs for cooking fuel of oustees after their arrival on the site and for the longer term, *i.e.*, three to four years hence."

For long term action, he recommended that a number of woodlots of sufficient size may be grown to supply the oustees with fuelwood on a sustainable basis on maturity. Oustees may be employed to maintain them. Further suitable fruit bearing trees may be planted along the road and pathways within colonies.

Since the woodlots would take time to come up, the Bank also suggested that firewood may be provided on a grant or subsidised basis as the majority of the oustees will neither be willing nor able to purchase firewood.

It is not a fact that the State Government are not aware of this great economic problem. As far back as in 1978, the Rehabilitation Advisory Committee in its first meeting dt. 25.7.78 deliberated about raising village forests near the proposed rehabilitation colonies. According to State Revenue Laws, 20 per cent of the effective area of a village should be reserved for "gochar", village forest, "busti", other communal purposes and for future development. Hence these provisions of law were kept in mind while preparing the plan for the rehabilitation colonies. In addition to this, the Rehabilitation Advisory Committee in its second meeting dt. 10.9.79 took decision to keep an additional area of 10 acres of forest in each rehabilitation colony untouched as a special case, just to assuage the local feeling.

In spite of all these pious decisions, the prevailing situation is bleak. The reason is not far to seek. There has been indiscriminate felling of trees by oustees as well as by the host population; without any replacement. Trees which were in existence are now extinct. Woodlots that would be raised would also meet the same fate. It is not that in apprehension of exploitation of tree growth, woodlots should not be grown. But there should be proper care to maintain them and replenish them too at appropriate time in a systematic manner. The Regional Plant Resource Centre, Bhubaneswar has done some research work on biomass potential of some selected fuelwood species. They have also identified some quick growing fuelwood species, which can yield fuel within one completed year. These species can be tried in the settlement colonies.

As regards threshing yard, the Bank team observed that cement drying/ threshing platforms should be provided to enable to oustees to process their grain crops without loss after harvesting. But this does not seem to be a sound proposition. The traditional practice of threshing in the countryside is 'khala', which is a threshing platform in the open level ground plastered with mud and cow dung. Cemented platform will cause the cattle slip off their hooves while threshing.

These are the realities experienced in the colonies.

Situation Outside the Settlement Colony

Though the Rehabilitation Organisation of Government have not taken any action so far to know about the oustees who moved on their own in different directions other than the settlement colonies, the author undertook a sample study in respect of few oustee families who have made their own resettlement arrangements within close proximity of the U.K. Project area. Table 6.17 gives an idea about their economic status prior to displacement as well as after displacement. The informations were collected in the same format as in Annexure-VI.

A close study of the Table 6.17 would indicate that after displacement, the economic standard of these people has gone down. Each family got lump-sum rehabilitation grant of Rs. 14,040 to rehabilitate himself/herself in the right way. But while 41 families had got in the old places, agricultural landed property of their own, of whatever, size it may be, in the new villages, only 14 families engaged themselves in cultivation. Prospects of employment opportunities in the project were not high for them although the number increased by four families. Thus, there was a substantial increase in the number of landless wage earning families. Obviously the rehabilitation money received by them was frittered away and they resorted to wage earning as daily labourers. This is indeed a very serious economic situation which should be watched by our planners, administrators, sociologists as well as society.

Suggestions

"Habitability is the result of the interaction of the natu-

Table 6.17
Economic Status of Few Settlers Outside the Colony

No.	village submerged.	No. of families displaced.	Population.	New village of settlement.	Families settled	Previous occupation	Present occupation	Present population.
1	2	3	4	5	6	7	8	9
1.	Koranga	322	N.A.	Nua Koranga.	19 (All SC)	(1) Cultivation-10 (2) Landless wage earning - 9	(1) NMR employee -2 (2) work charged employee -2 (3) Cultivation-5 (4) Landless wage earning-10	119

(Contd.)

Table-6.17 (contd.)
Economic Status of Few Settlers Outside the Colony

1	2	3	4	5	6	7	8	9
2.	Puki (Koraput P.S.)	113	800	(1) Nua Puki-2	S.T. -16 S.C. -3 Gen-5	(1) Culti-vation-19 (2) Landless wage earning-5	(1) Culti-vation -6 (2) Landless wage earning-18	120
				(2) Nua-Ganjeipadar	S.T.-7 S.C-5	Cultivation - 12	Landless wage earning-12	62
3.	Puki (Nandpur P.S.)	120	800	Nua-Puki-1	22 (All S.C)	(1) NMR employee -8 (2) Landless wage earning-14	(1) NMR employee-8 (2) Cultivation-3 (3) Landless wage-ear ning-11	140
Total	555	1600	S.T.-23 S.C. -49 Gen - 577			(1) Culti vation - 41 (2) Landless wage ear ring - 28 (3) NMR employee- 8	1) Cultivation - 14 (2) Landless wage ea-rning - 51 (3) NMR employee - 10 (4) Work charged employee -2	14 441 51 10 -2

ral factors of land fertility, climate, topography and the developments sought to be inscribed on it by man" says K.V. Sundaram in "Urban and Regional Planning in India" (a book published by Vikas Publishing House Pvt. Ltd., Delhi). To elucidate it he quotes Arthur Glikson, the Regional Planning and Development, as under:-

> "It is a state of balanced relationship which was to be achieved by man by careful selection, adaptation, planning and action."

In case of rehabilitation of Upper Kolab oustees, land selected does not suit them because of differential environment. The soil needs enough treatment for enriching its fertility; climate and topography are different because of differential altitude in comparison to the displaced area. Much is desired to be done in the developmental activities to provide minimum civic amenities. In other words, the implementation of the rehabilitation scheme in the field has failed to restore the social and economic condition of the oustees, least to speak about improvement if any as contemplated by the Government. There is lack of coordination in the activities of the resettlement staff and that of the resettlers. It may be put in nutshell that the "After-Care" treatment of the resettlers that is required of the officials, has been wanting very much. Every individual has some pride in himself or herself. But in case of the resettler there does not seem to be anything to feel proud. Socially he is looked down upon by others as a man of 'submerged area'; a "bastuhara". Economically he feels insecure. He feels himself diffident to merge into the mainstream of modern civilization. Sincere efforts are required to be made as follows to solve the problems:-

(a) Completing Irrigation Works

The agricultural lands allotted to the resettlers are mostly poor in quality. Only with the spread of irrigation facility from the U.K. Project, they could sustain the resettlers. Hence it is very much incumbent on the part of Government that the progress of work in this regard is accelerated and the colony lands are provided with irrigation on top-most priority.

(b) Substitude Irrigation Facilities

74 families (28 in colony No.5, 1 in the colony No.6 and 45 in colony No.7) out of 424 families resettled so far have been allotted non-irrigable land only, though the rest have got either irrigable land only or both irrigable and non-irrigable. Providing irrigation to the irrigable land will bring prosperi ty, whereas the resettlers with non-irrigable land will always look askance at the Rain-god. Possibility of constructing deep-bore lift points, irrigation tanks and water harvesting structures etc., may be explored.

(c) Distribution of 'Pattas'

'Pattas' may be distributed to all the resettlers on priority so that a sence of belongingness, right and title is reposed on them.

(d) Demarcation of Allotted Land

The lands allotted to the resettlers may be demarcated in the field and encroachment, if any, may be removed.

(e) Extending Credit Facilities

Credit facilities may be extended by Co-operative Banks/ Nationalised Banks for acquiring agricultural implements, fertilisers and pesticides etc.

(f) On-Farm Development Works

Irrigation system development and land consolidation should be done simultaneously. Although this, the on-farm development works should also be taken up. In rice-growing areas of South-Eastern countries, especially in Indonesia, the difficulties of acquisition of land for field channels and water courses are done away with as their alignment is done with the consent of all farmers concerned to part with required extent of land. Since irrigation would take time to come to the area, there is sufficient time to think in the above line.

(g) Co-operative Farming—Farmer's Association

Co-operative farming may be introduced. In the absence of regular banking structure, co-operatives might be able to provide an alternative source of agricultural finance. The author in course of his training in the U.S.A gathered that the Federal Bureau of Reclamation does not construct any water resources structures for irrigation unless and until the farmers come forward with an undertaking that they will be responsible for the construction and maintenance of all the channels, water courses, etc., at their expene. However, the Federal Bureau will provide technical advice if and when required. To that effect the farmers are required to form Associations between them. This system works very well.

The philosophy of water users' Association is yet to catch the imagination in Orissa. But such experiments in Kerala and Maharashtra have gained ground. Irrigation system development in U.K. Project is in initial stage. Efforts should be made from now on to form associations by the farmers in the irrigated area so that operation and maintenance of the irrigation system can be made over to the farmers. A question may arise as to for what benefits the farmers would shoulder such high responsibility? The answer is that, firstly, they will have the satisfaction and pride of belongingness. Secondly, the recurrence of overwatering in the head reaches and scarcity of water in the tail areas will not be there. To add to it, to promote greater awarness, incentives can be granted to farmers either in shape of reduction in payment of water rates or in awarding suitable packages of seeds and fertilisers or agricultural implements.

(h) Agricultural Development Package

The landless oustee and the landlord oustee have been placed in one category of land owners of equal status. It is not late to evolve a "development package," aiming at re-constructing the production base of the resettlers. Mere compensatory package would not be enough to raise the standard of living of resettlers. In order to assist the resettlers to further enhance their economic standard, family oriented development programmes *viz.,* land development and levelling, supply of modern agricultural implements, bullock carts, bullocks, goats and sheep, milch cows may be taken up intensively.

(i) Dairy Development

There are very few resettlers belonging to milman community. But it has almost become an economic hobby and necessity with everybody to take up dairy. A good number of cattleheads are there in colony No.7. Firstly, the livestock already existing in the colony should be provided with an explicit programme for pasture land development. Secondly, livestock development programme should be chalked out so as to provide traction animals as well as livestock to the farmers. Such facilities when provided, can create dairy market in Kotpad, Boriguma, Nowrangpur and as far as in Jeypore town which is fast developing into a cosmopolitan town.

(j) Plantation

The forest environment near the resettlement colony has already been disfigured to meet the fuel need of the resettlers as well as the host population. A programme to provide long term fodder and fuel requirements is, therefore, very much necessary. Sizeable area has been earmarked in each and every colony for future development of grazing ground. Tree, shrub and grass species which are useful to the resettlers can be grown in this area and the buffer zone so created can be handed over to the settlers for maintenance and management.

The State Government have got a scheme known as social forestry. The Orissa Plantation Development Corporation have also got the technical know-how to help in this regard. The borrow land along both sides of proposed canal embankments can also be brought under plantation. In addition to this, avenue plantation can be done along the road sides in the colonies. Fruit bearing trees may also be grown along the roadsides which would bring income to the settlers. The tribals are very adept in collection of resin, honey, fruits and nuts. Hence care need to be taken to grow trees which would yield these products. Thus, economic plantation and plantation of quick growing species should be raised to cater to the needs of fuel, fodder and timber. The plantation so raised will also help in checking soil erosion and in preserving the eco-system.

(k) Fuel Requirements

Creation of woodlots by way of plantation would take time. Hence arrangements are required to be made to meet the immediate fuel need of the settlers. The following are some of the propositions which can be tried:-

(*i*) The Orissa Forest Corporation can open fuelwood depots for selling fuelwood at subsidised rates.

(*ii*) The State Government have got schemes to subsidise installation of biogas plants. To minimise expenditure, instead of installing small and individual units, community units can be installed.

(*iii*) The settelment colonies are located in an area where the sky remains clear for about 8 months in a year. The State Government have got schemes to popularise use of solar cookers. These schemes can be popularised on subsidy basis.

(l) Labour

There is scope for the resettlers to earn wage by putting labour in the canal construction works. The project authorities should ensure that there is no resistance in this regard from the host population.

(m) Mass Employment Generation

Skill development training programmes may be arranged so that the resettlers can create self-employment opportunities. For this assistance from Ministry of Science and Technology, Government of India under mass employment generation programme can be availed of.

(n) Amity Between the Hosts and the Guests

An awareness is to be nurtured in the minds of host population that the resettlers are human beings like them and there should not be any differential identity. All social and cultural functions need to be organised without any discrimination.

(o) Drinking Water Facilities

The drinking water wells, inadequate in depth may be deepened and the excavated portion of the tanks may be enlarged to ensure long storage of water. Test drilling should be carried out during high summer when the water table is at its lowest, and bore holes with pumps are put in those locations identified to be having sufficient potable water.

The exercise in this regard may be an expensive proposition. But no government which believes in welfare State, can afford to neglect this essential aspect of human living.

(p) Schemes for Oustees, Still Struggling for Resettlement

The project authorities have no record to indicate the number of oustees who were given employment in the construction stage of the project. On the other hand very few people (25 in Class IV and 23 in N.M.R) were employed in the project. They were provided neither any vocational training facilities nor any self-employment schemes were floated for the benefits of the resettlers. About one-eighth of the displaced families have chosen to move to the resettlement colonies. About one-fourth seem to have made their own arrangements for rehabilitation in the true sense. They are the selected few who were economically better off. The rest who are mostly tribals and Scheduled Caste people and who had no assured means to sustain, appear to have eaten away the money received towards compensation and rehabilitation grant and selectted any and every place on the bald hills to the close proximity of the project site in the hope of getting employment on daily wage basis.

(*i*) They have eroded the bald hill slopes further which would cause soil erosion.

(*ii*) They bother the project authorities every now and then for employment, which is not in existence.

(*iii*) They resort to committing theft. Instances are not rare where some of the miscreants from out of these people have broken the roadside culverts-parapets, distance indicators, etc., to get iron roads and sell them at a paltry rate. This is how incidence of crime has risen.

Hence it is very urgent to make a census of the aforesaid category of displaced persons and motivate them to resort to gainful occupation. Here again appears plantation to be a good answer. Several entrepreneurs from Andhra Pradesh and Kerala have come down to Koraput district and started growing tree plantation in commercial scale.

Government machinery, if devoid of missionary zeal, fails in this regard. Hence either voluntary organisation may be induced to help these oustees to grow plantation on lease basis, or private entrepreneurs are allotted lease hold plots for plantation with stipulation that they would engage the oustees for maintenance of the plantation; on failure lease will be terminated. The climate and soil in the catchment area are ideal for growing lemon, oranges. Fruit bearing trees if grown will not only act against soil erosion, but will also open commercial avenues. Moreover the tribals once inducted to this occupation, will stick on to it for ever because; they might be practising 'podu' cultivation, but they know very well how to preserve fruit bearing trees. Their culture is very much interlinked with forests. They have got a great sense of belongingness for the forest and through ageold perception they are adept in selecting trees which are useful and unuseful. Hence our planners should identify the tribal leaders who can participate in the afforestation programmes and their management. This will not only help in ameliorating the local people especially the tribals in catchment area and their socio-economic development by providing them with continuous year-round gainful employment, but will also wean them away from the pernicious habit of "podu" cultivation.

(q) Spread of Education

To make the resettlers aware about the socio-economic developments that is necessary for them, Adult education schemes and non-formal education schemes may be introduced.

(r) Opportunities Outside Settlement Colonies

The economic opportunities outside the settlement colonies seem to be better. They are:-

(i) Reservoir Pisciculture

Reservoir pisciculture has great prospect in Orissa. Existing reservoirs of major and medium irrigation projects in the State have been transferred to the State Fishery Department so far as fishing rights are concerned. Of late, Government in Irrigation Department in collaboration with the Fishery Department and Central Institute of Fresh Water Acquaculture, Kausalyagang, Bhubaneswar have prepared a pilot scheme for floating net or cage fishing in reservoirs. The objectives of the scheme are:-

(*A*) to provide gainful employment to the affected population of the reservoirs.

(*B*) to utilise the large water body for commercial production of protein food.

(*C*) to develop small scale entrepreneurship.

(*D*) to develop techniques for rehabilitation of people affected by construction of reservoirs.

This scheme has been formulated on the line of technique adopted by the World Bank aided Saguling reservoir in

Indonesia. Financially it has also been calculated that three cages, each of 6 X 6 X 2 metres, three cages, each of 3 X 3 X 2 metres, two cages, each of 7 X 7 X 2 metres and two cages each of 10 X 10 X 2 metres size will support seven families at a total expense of Rs. 9,97,432/- annually (both recurring and non-recurring). There will be two catches every year and the net profit would be of the order of about Rs.1.4 lakh if fish is sold at an average price of Rs. 12 per Kg. Training and education in this regard will be provided by the State Fishery Department.

Experience in Saguling reservoir has further shown that one cage's profit in the reservoir is more than that of two acres of paddy land. Keeping the dearth of suitable paddy land in view, the oustees can be motivated to take up cage pisciculture in U.K. Reservoir and land for homestead purposes only can be allotted to them in the reservoir fringes so that the cage owners can be available all the time near the cage. In order to keep the reservoir water free from pollution, more than one per cent of the water surface should not be utilised for such pisciculture. Hence number of persons that can be engaged in cage pisciculture can be worked out.

There is great market potential for fish in Jeypore, Koraput, Sunabeda and Damanjodi in close proximity of the reservoir. Distant markets are Vijayanagaram, Visakhapatnam, Rayagada, Jagadalpur and even Bhubneswar. This type of pisciculture will also provide subsidiary occupation to those who could be engaged in manufacture of fish feed.

(*ii*) Sericulture

Sericulture through mulberry plantation has gained popularity in many parts of the State. It is said that sericulture can become a profitable enterprise only when irrigation is available. Normally there would be few crops in a year. In Karnataka it has been experimented that one acre of irrigated sericulture will give a net income of Rs.5,600 in the first year and about double the amount from second to fifth year. Irrigated agriculture from one acre of land cannot possibly bring better income than one acre of sericulture. It is desirable that the State Industries Department champions this scheme in right earnest in these colonies.

(iii) Poultry

Broilers and eggs are considered hygienically better substitutes to meat. The birds also grow quickly. There is market available not far off. The refuge of the birds can also be processed as fish feed for the cage pisciculture.

(iv) Nursery and Bee keeping

The land scape around the U.K. reservoir is very ideal for growing flower and fruit nurseries on commercial basis. This will also help in developing bee-keeping in the area. Resettlers can be motivated to take up these self-employment generating schemes.

The general position is that a sizeable proportion of the working force of the resettlers is engaged in the agricultural sector. The other non-agricultural operations are almost non-existent. This is not a healthy symptom. The capacity of the land, with the existing holding size, techniques of cultivation are inadequate to sustain the resettlers. This is obviously the result of not providing new avenues of employment through other probable sectors as discussed above. It is necessary that a harmonious development of all sectors is attempted.

SECTION-E : CULTURAL

Crisis in Development

With advancement of science and technology, there has been rapid improvement in economic, social as well as cultural status of human being. Government have also floated innumerable development schemes for the upliftment of Scheduled Caste and Scheduled Tribe people, keeping in view that they constitute about 40 per cent of the total population. But as in case of other projects, in Upper Kolab Project too, in the resettlement colonies there has been anti-government tendencies which can be traced to a crisis in development of their feelings of identity. The resettlers behave in a hostile manner towards the Government officials because they feel that the Government have left the resettlers yet to be fully resettled. The promises and assurances given by Government for leading a decent living still remains unfulfilled. The officials in-charge of rehabilitation activities do not pay visits to the colonies regularly; unless and until there is a crisis of failure of drinking water wells or some V.I.P. visits them. Legitimate demands if not fulfilled in due course of time, get piled up resulting in restlessness in the minds of the resettlers and administrative crisis in Government machinery. This is the reason for which any time any Government official visits the resettlers to enquire about their well-being, they come up with a charter of demands and complaints. During the time of the visit of the author alongwith several project officials, the resettlers complained about non-receipt of full quota of the rehabilitation cash grant. On scrutiny it was found to be not true. The resettlers were in arithmatical errors. But the officials did not seem to have explained the correct position to the resettlers. In matters of money, one gets swayed by instigations. The officials should have taken note of it and explained the position to the best satisfaction of each and every resettler. They also complained about poor yield of crops from the lands allotted to them and inadequate supply of potable water.

Thus, the resettlers are uneasy with the state of affairs prevailing in the colonies. They also do not see bright future, because the Government machinery, so far has not been able to convince them about the reasons for lapses and the resolving programmes in offing. They have become restless and hence want quick action.

Another great crisis noticed in the catchment area is the influx of oustees. Their migration especially that of the tribal oustees has resulted in clearance of forests for cultivation. Hence there is a threat to forest conservancy. Another problem is " Jhoru" cultivation or stream bed cultivation by the people in catchment area. Such cultivation increases the silt load factor in the reservoir. It also threatens the water courses by flash floods.

Striking a Balance

The resettlement colonies in Upper Kolab Project have been set up in existing revenue villages. As such under the law of the land, they do not have got any independent existence of their own. But strangely enough, the spirit of belonging to one village community still remains to develop both in the resettlers and in the old villagers. The resettlers seem to be treated by the host community as alien and lower in cultural status. There seems to be a lack of conscious effort on tha part of the Government machinery to foster a community spirit among both the hosts and the new population. Efforts are necessary to be made to strike a balance and remove the hardship to both the communities. This is possible through motivation by the Block Development Agency and Voluntary Organisations.

Ideology of Village Community

The new settlement colony is new in all respects and has got no comparison with the old village which is no longer in existence because of submergence. A traditional Indian village and more so an Orissan village is a very culturally compact unit containing people from different sect and community and who are dependent upon each other so far as socio-cultural activities are concerned. But the resettlement colony is more known as a habitation of group of households having no intimate link with the amenities available. There is meagre social security as the oustees have been scattered and shifted a different places. The sickness for village pond, village shrine and the burial ground is very much lacking as they are yet to develop. The tanks dug are devoid of water. Each old village had its own deity around whom the villagers before displacement used to gather on festive occasions. Modern civilization had affected their living very little. It was unthinkable for them to leave their villages which also meant leaving their deities. But once they knew that there was no alternative but to move in time before everything that they loved was submerged; the very organised community life helped them to evacuate the old sites in a very orderly manner. They did not put any resistance. But in the new sites they have not been able to restore their religious and cultural activities. However, a big leap which is noticed in their cultural advancement is that, although very much different in theme and essence, the resettlers in colony No.6 have adopted the ceremonies that are in vogue in nearby villages namely Patuguda, Dhanpur and Chitagaon. They perform "pousa Parba", "Chaiti Parba" and "Nua-Khai."

Suggestion

It has been generally experienced that the resettlement schemes are extremely stressful for the resettlers and this is more so in case of the tribals. Displacement of human settlement because of submergence causes disturbances which require cultural as well as social adjustments in the new settlement. On account of unequal level of income and dissimilar cultural background, the resettlers require enough time to merge into the new environment. This causes hardship to both resettlers and host community. Sincere efforts are necessary to remove this cultural imbalance. In this connection it may be necessary to quote W. Partridge, Resettlement Specialist of the World Bank from his "suggested outline and recommendations for an action plan for resettlement and rehabilitation in the Upper Indravati Project" in Orissa as under:

> "In order to reduce high levels of stress in resettlement most settlers become temporarily much more culturally conservative than is usual—they will avoid new and unknown activities and the elements of risk that such always entail. This means that for the first several years the development plan should include a transition period when settlers are able to continue traditional cultivation practices, cropping patterns, crop selection, technology, and so forth. Innovations should not be forced so soon, although these will eventually

be required to make the new farm economically viable. I am aware of no project where innovations have been successful sooner than 2 years following transition. On the other hand, if such changes are made after the settlers have become established in the new sites and achieved a subsistence level income from their new lands global experience show them to be even more receptive to innovations than ordinary farmers. The development plan for the New Settlements should show an implementation time table which reflects this lesson."

Secondly, detailed anthropological studies to assess the displacement of tribals due to construction of U.K. Project and their life style should be taken up. The thrust is that tribal culture in brink of disintegration should be conserved as far as practicable. Determined efforts are necessary to prepare suitable plans and programmes for the improvement of the quality of tribal environment.

SECTION-'F' : HUMAN

The Lay-out

All the resettlement colonies have been laid out on the same plan. Drawing No.7 is the index plan showing all the resettlement colonies programmed by the project authorities. Drawing No.8 is a true representative of colony No.3. It would be seen that the plan has been made to accommodate the present requirements and areas have been kept identified for future development. However, colony No.3 has not been rehabilitated by the project oustees.

The position in the field in November 1989 presented a desolate appearance. Paddy crop had been reaped and patches of 'niger' crop only was here and there. The tanks were dry and the open wells were about to dry. Hence for want of water, kitchen gardening was almost non-existent. The only way of greenery was writ large in the castor plants and few drumstick trees. Whatever financial value the newly harvested crop might bring to the resettlers; a lady member of them was zooming with country liquor in open day light. This is perhaps the only form of aristocracy and satisfaction, the resettler could exhibit.

Psychology of the Resettler

Experience in Rengali Project shows that the oustees were very resolute in their stand not to evacuate. They formed strong Committees to oppooe to the construction of the dam. But in case of Upper Kolab Project there was not much of resistance. This is primarily because of the fact that the oustees mostly belong to weaker section of the society and the peoples' representatives co-operated with the Government machinery. Hence the initial reaction of the resettler was not very violent. On account of his simplicity, he was not very much apprehensive as to what would happen to him in the new environment. He had absolutely no idea as to what would be the colony like. Rather he cherished high hopes about his colony life as was given to him by the project authorities. But the new life in the colony amidst all odds has made him psychologically upset. Delay in getting patta and irrigation, difficulties in getting potable water and lack of brotherly feeling with the host population have disheartened him. The growing dissatisfaction among the resettlers has also created a sense of frustration among them.

Pressure Tactics

Study revealed that comparatively poorer class of people from among the oustees have been shifted to the resettlement colonies. This is mostly due to the facts that, (*a*) the comparatively well-to-do oustees who got better compensation money could make their own rehabilitation arrangements, and (*b*) the poorer people were more prone to the pressure and compulsions to move to the resettlement colonies.

The latter category of people took time to adjust in the environment which is devoid of employment prospects to fall back upon during the transition period between shifting and resettlement.

Desertion

It has been witnessed that oustees from the plains resettle in new colonies in quicker time than the oustees from the hills. The latter are carried away by receipt of handsome compensation money and hence are always tempted to desert the new settlement in search of better fortune elsewhere. This is conspicuous in case of colony No.4 (4B) where all the 227 families of old village Parojapondi were given accommodation. Parojapondi is a marooned village and on special consideration, government agreed for their shifting. But in reality, 97 families from out of 227 families, said to have shifted were not staying in the colony at the time of the author's visit on 15.11.89. It was gathered that they were still having a stake at the old site and thereby enjoying double benefits. The project authorities did not seem to have paid any attention towards this. Any delay in taking corrective action in this regard is likely to tempt other similar resettlers to desert the colony.

Another important factor leading to desertion, is the attitude of the host population towards the resettlers; which should not be lost sight of. The author worked for sometime as the Sub Collector, Malkangiri in Koraput district, at a time when the Dandakaranya Development Authority was in its height of activities. From personal experience it was known that some of the local tribes men were trying to adopt modern methods of agriculture after coming in contact with the refugee resettlers from East Pakistan (Bangladesh). They took up 'Mesta' (Jute) cultivation. Some of the 'Koyas' (who even today wear a 'Kaupin' only) took to wearing of dhotis and participated in the festivals of the resettlers, like Durge Puja. Thus, the local people got some indirect benefits. But instances were also not rare when these resettlers induced the local tribals to part with their agricultural yields during harvest at a very low price. During lean months when the tribals ran short of food, the resettlers also granted loans at a very high rate of interest. It was not surprising to note that tribal land also gradually passed into the hands of the refugee resettlers.

The resettlers of Upper Kolab Project colonies, had a rich harvest of niger crop during 1988-89. Outside middlemen did a handsome profit in niger trade. It is the unscrupulous tradesmen who know the art of trapping the resettlers to get loan facilities, mortagaging land, etc. The author got such a hint also in course of interview with the resettlers. It is, therefore, very much necessary that adequate safeguards are provided to protect the resettlers from exploitation of this kind; or else further desertion in compelling circumstances cannot be ruled out.

Responsibility of Society

The oustees make a sacrifice for the nation in general and several beneficiaries especially in the Command Area in particular. Hardly there is any river valley project in India where resettlement of the oustees has become successful without any stigma. They really deserve to be treated well. The art in this regard cannot be taught. A sense of oneness can only help the situation improve for the future resettlement programmes to be implemented.

Suggestions

The resettlers make a sacrifice without which neither there would have been additional power generation, nor creation of additional irrigation potential. Their contribution to the economic upliftment of the State is, therefore, more than that of any other citizen. The sufferings borne by them should, therefore, be adequately compensated by making their living condition in the new environment more attractive. If adequate care is not taken to deal with the human aspects of the problem, by necessity, they may assume political overtones making the successful implementation of the programme more difficult and complicated. The oustees should be made aware that there is somebody in the resettlement implementation Agency to feel for them and work for them in a missionary zeal, so that they are not made to run from pillar to post. Voluntary organisations should be actively associated to cultivate this culture so that atleast no further desertion occurs.

The author, although is a student of Geography, professionally is a bureaucrat. His practical experience in administration leads him to believe that, administrative measures alone will not help in resolving the problems relating to displacement and resettlement. Whether one is a generalist or technocrat in Government employment, he is primarily a bureaucrat and is always anxious to see that Government decisions whether right or wrong are implemented. The bureaucrat sees only the administrative part to enforce policy decisions of Government. Eviction and finding out alternative site for settlement are his responsibility. But it should be considered that there are many social as well as sociological factors too. Effective resettlement and rehabilitation depends on the Agency which is alive to the realities around it and which knows the feelings of the oustees. It is here that the role of sociologists, voluntary organisations comes to play. They are required to be involved so that an atmosphere is created in the minds of the oustees that all that is being done, is being done for a greater cause and the Government are really serious about the well-being of the oustees.

The author is tempted to quote here Pandit Jawaharlal Nehru, the creator of modern India:

> "Planning is not merely a question of drawing up a list of projects and finding resources to execute them. These two tasks, no doubt, form the base of planning.
>
> Among the resources, thought is mostly given to the financial aspects and in fact, there is the danger of this aspect, rather over-shadowing and suppressing the other aspects. The financial aspect is important but is far less important than the human aspect."

SECTION-G : FINANCIAL

Budget Provision and Expenditure

Upper Kolab Project is a State Government Project. Smooth evacuation of displaced families and their effective resettlement are the responsibility of the State Government. In keeping with the rules in force, all expenditure incurred towards land acquisition, payment of compensation and rehabilitation etc., are incorporated in the project estimate. The statement in Annexure V is an obstract sheet of the revised estimate of the sub-head "Reservoir" updated in 1987 under which the expenditure is met. All the expenditure incurred through the Spl. Land Acquisition and Resettlement Officer are towards compensation for land acquisition and payment of rehabilitation cash grant. All expenditure incurred through the Executive Engineer, Upper Kolab Development Division are towards survey and investigation, F.R.L. marking, reclamation of the land earmarked for settlement colonies, providing common facilities there, etc.

The Table 6.18 gives an idea about the year-wise budget provision and expenditure incurred by the Spl. L.A. and R.O. and the Executive Engineer, Upper Kolab Development Division.

Table 6.18

Year-wise Budget Provision and Expenditure (in Rs. lakh).

Year	*Expenditure incurred by Spl. L. A & R.O.*	*Expenditure incurred by E.E., U.K. Dev. Division.*	*Total*
1976-77	—	0.85	0.85
1977-78	—	59.79	59.79
1978-79	—	25.00	25.00
1979-80	—	126.15	126.15
1980-81	—	118.83	118.83
1981-82	30.84	295.21	335.05
1982-83	6.99	62.58	69.57
1983-84	0.21	44.84	45.05
1984-85	94.79	129.81	224.60
1985-86	300.75	43.08	343.83
1986-87	169.79	231.03	400.82
1987-88	227.10	48.08	275.18
1988-89	N.A.	N.A.	123.85
1989-90	50.99		50.99
	Total 890.46	1185.25	2199.56

Difficulties

Land acquisition work has been completed in respect of the area of displacement. Shifting work relating to 104 families only was left to be done in 1991. But the civic amenities provided in the colonies are not in satisfactory condition. A detailed assessment is necessary to find out the financial involvement towards deepening of tanks and wells, rectifying the tubewells and providing additional deep bore wells, repairing the link roads, electrification, etc. The Orissa Bridge and Construction Corporation, entrusted with the work of realignment of N.H.43 and M.D.R. 55 (Pithaguda bridge and approaches) have not completed the work so far even after extension of time from year to year. This has resulted in cost over run. The project (dam) is in its last stage of construction and irrigation system are an altogether separate project which does not include rehabilitation aspects of the oustees of the U.K. Dam Project. From personal experience it is learnt that provision once not included in the administratively approved estimate of the project makes it difficult to get funds for the future needs. It is unfortunate that there are still several unsettled land acquisition claims in respect of Hirakud oustees although the project was completed more than 30 years ago. Construction of Rengali Dam was completed and the first unit of power generation was commissioned in 1985; but maintenance of the civic amentities provided there has become a very costly proposition and for which there is no fund available.

All these uneasy state of affairs are the result of lack of systematic planning and preparation of proper estimate. In their anxiety to get a project cleared, Government machinery underestimate the loss due to submergence and cost of resettlement. Resettlement schemes should always be designed to help the displaced persons in regaining at least, if not improving their previous standard of living. Hence various resettlement options should be planned and cost analysis prepaed to be reflected in the budget.

Suggestion

What can be done to reduce both financial needs and financial constraints ?.

Very often based on imperfect and inadequate survey and in anxiety to get a project cleared quickly, project reports as well as estimates for them are prepared. The time schedule for completion of the project construction is also not based on perfect analysis. Consequently there is cost and time over run. In case of Upper Kolab Project, it is very much necessary to make a stock-taking of all the incomplete items of works and of those essential additional works and prepare a firm cost estimate with time frame for completion in all respects. Whatever constraints there may be to get funds, provision has to be made to fulfill the requirements.

SECTION-H : ORGANISATIONAL

Structure

In Chapter 5 the organisational set up in-charge of land acquisition and rehabilitation activities has been discussed in brief. It would show that all these activities are in hands of Government officials. They form two distinct organisations, *i.e.,* the officers organisations and the staff organisation. In the officers class come the Zone Officers, Land Acquisition and Resettlement Officer from the Revenue Department and the Junior Engineer, Asst. Engineer and Asst. Executive Engineer/Executive Engineer on the construction side from Irrigation Department. The staff constitute the revenue supervisor, revenue inspectors, Amins, Chainmen and Work Sarkars etc. Although some line is maintained for the engineering personnel, there is no line for the revenue personnel. While the Zone Officer can be posted as Tehsildars, Block Development Officer, etc., on administrative line; the Land Acquisition Officer can be posted as Addl. District Magistrate in a district or Deputy Secretary to the Government in any Secretariat Department. Thus, these functionaries are interchangeable. This results in lack of scope for expertisation in the field of land acquisition and rehabilitation works. On the other hand, the engineering personnel remain as engineering personnel irrespective of their posting.

The Chief Engineer in-charge of the Project is the functionary head of the project, under whose guidance all the engineering personnel work. On the other hand the Collector of the district is the immediate administrative head of the Spl. L.A. and R.O. and all his subordinates. For anything and everything the Spl. L.A. and R.O. waits for the instructions of the Collector. The vertical heirarchy of administration above the Collector is the R.D.C. and the Government.

Little Regard for Public Agencies

In Government machinery, an employee, especially a generalist claims to be a master of all trades. It happens so because, the existing system wants him to behave so. He also does not want any sort of interference even if it is constructive in nature. The aversion is mostly due to the fact that the public in general are very critical in their approach to Government machinery. There has been a feeling everywhere that Government should do everything good for the public. Construction of projects, their operation and maintenance are the responsibility of the Government. In the beginning of the planning era, there was a share attached to each and every community development scheme for peoples' participation. But with passage of time, no importance was attached to this excepting indicating a percentage on paper only. Consequently lapses committed, if any, in the scheme are magnified and criticised vehimently. Instead of becoming tolerant to criticisms, Government machinery also acts otherwise and takes care not to associate the public/non-government organisations in the implementation of schemes. This is not desirable.

Avoidable Hardship

In the prevailing organisational set up, the revenue officers are incharge of land acquisition and rehabilitation works and the infrastructure development works in the resettlement colonies are the responsibility of engineering personnel. Unless very close co-ordination in the activities of both these wings is maintained, no success can be achieved.

From personal experience of the author it is mentioned here that the system is faulty. There has been mudslinging by one wing against the other. No doubt, the engineering wing files requisition with the Land Acquisition Officer for acquisition of land. But in umpteen number of times the latter returns them to the former with queries and objections in a routine fashion without application of mind. While the engineering personnel shift the responsibility of evacuation of displaced persons and their rehabilitation in colonies or elsewhere as the case may be, the revenue personnel in-charge of resettlement and rehabilitation blames the engineering personnel for lapses, if any, in the infrastrcture development works. In these matters, the role of the Chief Engineer who is the head of the project administration has not been defined in clear terms.

The five organisations which are concerned with land acquisition and resettlement/rehabilitation are the Spl. L.A and R.O., Chief Engineer, Collector, Revenue Divisional Commissioner and the Government. The Government may come and go with the change of political power. The Chief Engineer is purely functional in character. His tenure, work programme are time bound. But the other three functionaries are statutory. It is desirable that they act in real sense of the term. In creating infrastructure development and civic amenities, the Chief Engineer depends on outside Government agencies *viz.*, Public Health Organisation, Electrical Department, Forest Department, Agriculture Department, etc. Generally it has been felt that the issues related to these agencies are left hanging between the scheme implementation and the agencies. This not only results in time and cost over-run, but also causes avoidable hardship to the resettlers. The Collector as the supreme commander of the district administration and the R.D.C. as the co-ordinating agency can do a lot to avoid these hardships.

Suggestion

A question, therefore, arises as to what would seem to be the best arrangement to ensure smooth and efficient resettlement of the oustees? The following answers may perhaps meet the query to some extent.

Splitting up of Existing System

The existing set up of Land Acquisition and Rehabilitation/resettlement responsibility may be split into two different organisations; (i) Land Acquisition and (ii) Resettlement and Rehabilitation; completely independent of each other.

It has been argued in many quarters that the work relating to land acquisition may be left with Revenue officials; but resettlement and rehabilitation activities may be given to officers of other disciplines. The World Bank have recommended in respect of projects aided by them in Orissa, that officers who are adept in social process of resettlement and development of new lands for agriculture should be entrusted with the work of resettlement operation.

The author does not subscribe to this hypothesis, because there is no other functionary than Revenue Officers in the State who are better oriented in a manner to meet the situation appropriately. However, some reorganisation is suggested as under:-

(a) Separate Establishment

Land acquisition is a subject of State Revenue Department. The Revenue Officer posted as L.A. and R.O. in the project is shown on deputation to Irrigation Department. But the paradox is that, although there is a Chief Engineer in the project and who is sufficiently senior in rank and salary to the L.A. and R.O., he does not function as the administrative head of the L.A and R.O. Organisation. This results in divided responsibility in the works relating to both land acquisition and resettlement. For that matter, the Irrigation Department has also not got much of jurisdiction, excepting acting as the administrative department; because the respective Collector and the Revenue Divisional Commissioner are the disciplinary ladder for the L.A. and R.O. On the Other hand, the Collector as well as the R.D.C., in the midst of their busy schedule hardly make time to supervise and review the work of the L.A and R.O. For effective implementation of L.A and Rehabilitation works, it is suggested that a full fledged

Directorate is set up under the charge of senior Administrative Officer as Director. He may be responsible for effective and smooth land acquisition and R and R activities in respect of all the development projects in the State or one Director exclusively for the irrigation and hydro projects can also be detailed keeping the work load in view. Duly assisted by other officers and staff he will supervise the work relating to land acquisition and rehabilitation of oustees of all concerned projects. The Land Acquisition Officer as well as the Resettlement and Rehabilitation Officer should be authorised to function as Heads of Offices with separate budget provision to operate. Such a provision will help him reposing confidence in him in all financial matters too.

(b) Development Division vis-a-vis Rehabilitation Organisation

For better co-ordination, the Development Division entrusted with the infrastructure development works may function within the Resettlement and Rehabilitation set up. In addition to this, Government agencies in-charge of Agriculture, Animal Husbandry, Fisheries, Soil Conservation and Forestry, Public Health etc., may be involved actively so that priority items are identified well in time and implemented within schedule. In order to achieve this, a very senior officer may be posted as Resettlement and Rehabilitation Officer. He should be made the Agency to help the resettlers for raising their standard of living by generating employment opportunities in agriculture, forestry, fisheries, cottage industry and by implementing all the development programmes of the State Government which are otherwise the responsibility of normal development agencies like D.R.D.A.,I.T.D.A., etc. For that matter he can be declared as Additional Project Officer D.R.D.A./ I.T.D.A. for limited jurisdiction of resettlement colonies.

(c) Role of District Administration

The District Administration play a vital role in the land acquisition and resettlement/rehabilitation activities. The Collector in his multifarious responsibilities hardly gets time to oversee these activities. The next officer to the Collector in the line is the Addl. District Magistrate. He deputises the Collector in statutory matters. Hence the Land Acquisition Officer and also the Resettlement and Rehabilitation Officer can be declared as ex-officio Addl. Dist. Magistrates in their respective jurisdiction for limited purposes.

They should be independent of the Chief Engineer in implementing the land acquisition and resettlement/rehabilitation functions.

(d) Training

Since resettlement and rehabilitation works are complex and difficult processes; the officers and staff entrusted with them should be trained suitably.

(e) Engagement of Outside Organisation

Although the Development Division looks after the infrastructure development works in the colonies, its activities are mostly confined to civil construction works like roads, buildings, wells and tanks. But it has to depend upon outside organisations for land reclamations, tubewells boring, plantation, etc. As an alternative it can be tried to engage an outside Agency especially public undertaking like Industrial Infrastructure Development Corporation which has got better expertise to deliver the goods. All the development activities can be brought under one package for execution. The Development Division of the Project may only act as supervisory body.

(f) Voluntary Organisation

Another great factor which leads to the successful implementation of resettlement scheme is the involvement of voluntary organisations. Some people call these voluntary organisations activists. But not that all such activists organisations smell a rat in everything. These bodies may be knowing the local population, their language and culture and be better suited to mobilise them to achieve the desired goal. These organisations also work as messengers of Government sponsored development programmes. Of late organisations like WIDA, AGRAGAMI and BANABHARATI have become pioneer organisations in this regard in Orissa.

(g) Creation of Authority

In Government machinery, works get delayed in course of processing them. There is delay in sanction of estimate, approval of tender, etc. Such delay is detrimental to the progress of project implementation. Delay in allocation of funds to the desired quantum also delays execution. These difficulties can be overcome by delegating adequate financial as well as administrative power to the Chief Engineer. Still a better alternative is creation of an Authority for each and every multi-purpose project to solve these problems at their level instead of referring them to Government for aid and advice.

(h) Participation of Resettlers and Host Population

It should be ensured that there is active participation of the displaced and the host communities in each and every phase of resettlement activity. Lack of such participation in Rengali Project created innumerable difficulties not only in timely completion of the project, but also in implementation of the resettlement scheme. The scheme is aimed at ameliorating the living condition of the resettlers. The host population is also likely to get indirect benefits from the schemes. Hence their participation will improve not only in understanding their needs, resources and preferences, but will also prevent avoidable mistakes and will smoothen the task of shifting of the displaced families and their effective resettlement. Serious conflicts are likely to occur between the host population and the resettlers on the ground that an artificial barrier is created between them due to establishment of better living facilities for the latter without any regard to such facilities available to the host population. In fitness of things services like communication, education, health and drinking water as far as practicable should be made available for both the groups of population. It should be made aware to both the communities that the assets created are meant for them and it is their responsibility to maintain them in healthy condition.

In order to achieve this, it may be necessary also to constitute Resettlement Committees at the colony level. The Committees would be associated with selection of sites for locating buildings, wells and tanks, preparation of land for cultivation, etc.

(i) Monitoring

Shifting of the displaced persons from the submergence area has been the main task of the project authorities. Equal importance should have also been paid to their effective resettlement. Some amount of sincerity seems to be lacking in distinguishing involuntary displacement from voluntary displacement. In the case of latter which normally happens in emergencies like war, natural calamities, etc., the people displace themselves without waiting for any direction and assistance to come. They also resettle themselves as they like; may be or may be not with any direction and assistance. But in the case of involuntary displacement, there is a sense of compulsion. It becomes mandatory on the part of Government machinery to ensure safe and steady shifting as well as resettlement. However, sound the scheme in this regard may be, its success depends on efficient implementation. It is, therefore, necessary that the implementation part of the work is monitored regularly by the project authorities at the project level and by the proposed Director, at the Government level.

(j) Evaluation

This is a very important task which should be entrusted to an impartial and outside agency, like a University or a social body which does not have any professional link with the project. The Agency should have considerable expertise and background knowledge in evaluating work. Further the evaluation aspect should not be confined to the progress of resettlement of oustees in the project sponsored colonies only. About 84 per cent of the oustees have moved on their own for unknown destination. Data relating to their whereabouts, social and economic status, etc. should be collected and all possible help be rendered for their resettlement if not done by that time.

In this context it is worth mentioning the role of the Rehabilitation Advisory Committee. As the situation prevails it is mostly a body which makes recommendations only. In twelve years since its constitution in 1978, it has met twelve times only, although as per the relevant resolution it has to meet at least once in three months. The body does not get the services of experts from the disciplines relating to agriculture, animal husbandry and horticulture which are very much necessary in case of resettlement of oustees who have been displaced from forest environment and resettled in an area where both irrigated agriculture and dry farming are necessary. There are elected peoples' representatives in the Committee, but the absence of any social scientist is very conspicuous. To ensure that there is no conflict of economic programmes with the social surroundings, there may be a social scientist taken in the Committee. The Committee should be functional as well as operational in true sense.

A word about the majority oustees who left for unknown destination. It is the responsibility of the government to see that each and every displaced family is rehabilitated successfully. Payment of cash grant in lieu of land does not absolve the government of its responsibility in this regard. Sincere efforts should be made to locate them, maintain their history sheets so as to enable them to get the development benefits that would have accured to them had they not been dislodged from their old habitation. Whatever perfunctory arrangements might there be, the resettlers in settlement colonies have been provided with certain benefits that would accrue to them as soon as irrigation water is available. But these oustees who left for unidentified places on their own have sacrificed a lot without being assured of any benefit from the project. Since quite a sizeable displaced persons seem to be residing in the catchment area, it is very urgent to prepare some action plan to safeguard their interest. The plan may contain programmes for (*i*) improvement of communication system, (*ii*) opening of adequate number of health care centres, educational institutions, (*iii*) development of pisciculture and dairy farming and (*iv*) suitable cropping pattern, etc.

SECTION-I : CONCLUSION

R and R a Sensitive Issue

Resettlement and Rehabilitation of Project displaced persons is a very sensitive as well as burning issue today. It not only touches the most weaker section of the community, but also brings in a vast change in the physical, social, cultural as well as economic environment of the affected population. Hence there is the need to strike a balance in a very systematic and effective manner. It is indeed a difficult task.

Objectives of R and R

The main objectives of the Resettlement and Rehabilitation should be:-

(*a*) to improve or at least regain the standard of living which the oustees had enjoyed prior to displacement, through the restoration of their productive capacity.

(*b*) to relocate the oustees as village units, village sections or families in accordance with their preference.

(*c*) to see that the oustees are fully integrated into the community to which they are resettled; and

(*d*) to provide the oustees with appropriate compensation and adequate social and physical rehabilitation infrastructures, including community services and facilities.

Observations by World Bank on R and R Works

The World Bank, in the past, have financed a good number of medium irrigation projects in Orissa involving human displacement. Presently they are financing the Upper Indravati Hydro-power Project. The Subarnarekha Irrigation Project is also in the pipeline for Bank assistance. From their experience in innumerable projects in India and elsewhere, they have observed;

"the failure, during project preparation, to carry out social surveys of those to be dislocated as well as of the host-area populations, combined with weak preparation of viable redevelopment alternatives, can make it impossible to conduct an adequate appraisal of resettlement plans, costs and organisational arrangements and can result in resettlement components which are under-designed, underfunded and understaffed. Inadequate concern with the severe consequences of involuntary dislocation or inability for preventing destitution—for example, by re-establishing those evicted on an alternative productive base — can defeat on important purpose of the overall development effort and of assistance by the Bank."

Further after carrying out a retrospective analysis of resettlement experiences in 1986 on a large number of projects, they have identified the following key areas which require immediate and considerable strengthening:-

> "first, the quality of borrowers' preparation and detailed planning of resettlement components must improve radically with respect to their economic, technical, sociological and organisational content; without comprehensive, detailed, and timely feasibility studies and relocation design, there is little that can be meaningfully done, at appraisal time and little that can subsequently guide implementation on an effective path;
>
> Second, increased attention must be given economically and socially viable options for developing the productive capacity of displaced populations through project—financed land—and employment based strategies;
>
> third; supervision by the Bank of the implementation of resettlement operations must be exercised more regularly, professionally and firmly, to help improve the performance of the agencies executing resettlement and ensure consistency of implementation with policies and legal loan agreements."

(*Source* - Michael M. Gernea in "Involuntary Resettlement in Development Projects" - policy guidlines in World Bank- financed projects World Bank Technical Paper Number 80).

Findings on R and R Works

In this Chapter experiences of some of the major river valley projects have been discussed. They are nothing but bitter. But it is worthwhile to quote Dr. Ethiraj Kumar of the Department of Anthropology, Andhra University, Waltair from his article on "Role of Enumeration on the Rehabilitation of the displaced due to Visakha Steel Project" as under:

> "But the findings of various social scientists point out that in many a case, the rehabilitation measures are 'mechanical' and 'inadequate'; exposing the affected population to deeprooted changes in life patterns and prolonged suffering. All along, the affected populations, mostly rural and tribal, were observed to bear with the situation socially due to their intrinsic inadequacies, *viz.,* poor literacy levels, ignorance of bureaucratic procedure, submission to authority, gullibility, etc. However, in recent times organised resistances against establishment of large projects is gaining ground, drawing from the distressing experiences of the displaced in earlier situations."

Story of U.K. Project

The story of Upper Kolab Project is not very rosy. The reservoir was impounded in 1986. But shifting of oustees was completed in 1991 only. Whatever temporary allurement may be there in receipt of rehabilitation cash grant, the arrangements made in the resettlement colonies attracted only a very small percentage (16.65 per cent) of displaced families to move there. The remaining families left on their own for unknown destinations and for whom no records have been maintained to watch their productive rehabilitation, if any.

The communication system disfigured due to submergence has not been restored completely. Public institutions submerged have not been relocated so far at substitute places.

The land based economy designed for the resettlement colonies has not been streamlined. Irrigation is yet to come. There is no programme also to provide any irrigation facility to those lands which cannot get irrigation water from the U.K. Canal System. Resettlers are still waiting for the distribution of 'patta' in their favour which would vest right and title on them to avail of credit facilities. The erstwhile dwellers of forests are digging roots of unrooted stumps in the reclaimed lands to meet their fuelwood requirement. No effort has also been made so far by the project authorities to meet the fuel requirements of the resettlers. There is no systematic cropping pattern and use of improved varieties of seeds, agricultural implements, fertilisers and pesticides has not become popular. Neither any vocational training facilities have been attempted for the resettlers, nor any employment opportunities have been created. Percentage of representation of oustees in the employment of project is utterly negligible. Least to speak about any self-employment generating schemes. There is also no family oriented development programmes for the resettled families in keeping with the changed environment to improve their economic condition. Catchment area treatment seems to be very haphazard. To restore the forest environment in the catchment area of the project, a comprehensive soil conservation scheme has been prepared. But it is yet to be implemented. The Environmental Management Scheme of Upper Kolab catchment with Norwegian Government financial assistance as projected by the Orissa Remote Sensing Application Centre if implemented, would go a long way in striking a balance. The civic amenities provided in the colonies are not satisfactory. They are not only inadequate, but also perfunctory. There is complete lack of maintenance and supervision. Water for the resettlers as well as for the domestic animals is insufficient. There is no plan for electrification. Community life is yet to develop. Much is desired to be done to establish amity between the resettlers and the host population. As a matter of fact no attention has been paid as yet to ensure that the host population in the neighbouring areas also get civic facilities as are available to the resettlers. This has resulted in growth of an unhealthy socio-economic atmosphere between both the communities. The result is that the resettlers do not get employment as wage earners in canal works due to opposition of host community. The old habitation in the catchment area has disappeared and lost under Upper Kolab reservoir. Those habitations, not submerged, now look lonely and depressed. Thus, there is change in ecological setting both in the catchment area and command Area. These changes should be taken into consideration and there should be planning and development in both the areas.

The Rehabilitation Advisory Committee which is the only organisation to review and monitor the progress of resettlement and rehabilitation activities met very infrequently and as the proceedings of the meetings of the Committee reveal much has not been discussed to either recommend improvements with reference to work done in the field or to resolve the problems touching upon the basic minimum requirements of colony life. The project supervisory officers in-charge of different activities in the colonies took their own time to ensure their operation. Consequently the resettlers elected to resolve to their fate. There was also nobody to guide the resettlers to spend the compensation money as well as the rehabilitation cash fruitfully. Involvement of voluntary organisation which would have gone a long way in straightening the issues has not been thought of. Cost expenditure to meet the lapses and additional requirements has not been worked out so far. Hence difficulties in making budgetary provision in this regard cannot be ruled out because, the project has been substantially completed.

In brief, the resettlement and rehabilitation programme in Upper Kolab Project has not achieved the desired objective. The organisation incharge of implementation of the same seem to have accepted it as a matter of routine; devoid of any real and mission. The main objective of the human resettlement as envisaged by the State Government is planning for an area of pioneer peasant colonisation and development. It is no gainsaying that the sites selected for human resettlement was almost a virgin area. However, bad in quality, the top soil may be, it offered unfettered freedom of choice to the project authorities to build ideal settlement colonies to suit the environmental conditions, economy and host society of the area. No doubt, the colony planned has been able to some extent in securing equity among its inhabitants in the means of production *i.e.*, the land. Equitable allotment of land is the basic guiding factor in this regard. But the scheme as implemented in the field has failed to provide the minimum standard of civic amenities.

The resettlers still look upto Jeypore for various socio-economic functions and specialised services. For their daily needs, they are dependent upon Kotpad or Sasahandi. The latter two are collection and distribution centres for the surrounding area and provide the basic services. They virtually function as the "nerve centres" for the colonies. It is, therefore necessary that due attention is paid to develop these nerve centres too suitably so that they attain functional self-sufficiencies. It would also help ensuring a very successful co-ordination of the rural planning function and in restructuring rural regions.

Prior to the enactment of the Forest (Conservation) Act, 1980, most of the resettlement issues were meted out as the government machinery considered proper. The Act brought some new dimensions especially relating to environmental changes due to construction of large reservoirs. The construction of Sardar Sarovar Project in Gujarat and Tehri Dam Project in Uttar Pradesh stirred the movement so much so that at a particular point of time it became impossible to see any project through by Government of India so far as forest clearance, environmental impact assessment and more particularly resettlement and rehabilitation measures to be taken in respect of displaced persons are concerned. Hence, it was advocated by many to go in for smaller projects (minor irrigation). But be it as it may, the large and multi-purpose projects have got their inherent advantages. They are also necessary and are required to be constructed. Human displacement is inevitable. But care has to be taken to plan minimum displacement. Resettlement and rehabilitation activities have to be pursued vigorously in a systematic manner within a positive time frame so that there is not much scope for debate. In succeeding paragraphs there are a set of guidlines suggested in this regard which may be tried.

Suggested Guidelines for Resettlement and Rehabilitation in Water Resources Projects

Definitions

(*a*) *'Water Resources Project'* is one which aims at providing water for irrigation, industrial and municipal use, generating hydropower and controlling floods etc.

(*b*) *"Land"* - The expression "land" shall have the same meaning as defined in the Land Acquisition Act, 1894, which states "expression 'land' includes benefits to arise out of land and things attached to the earth or permanently fastened to anything attached to the earth."

(*c*) *"Project Affected Person"* means a person who is affected in respect of his/her land including homestead and structures thereon due to construction of a Project.

(*d*) *"Displaced Person"* or "Oustee" means any person who comes under any of the following categories and belongs to the area which would come under submergence due to construction of any project. Any person substantially affected by project infrastruc ture works/ construction of irrigation canal system may also be taken in this category.

(*i*) Whose land to the extent of 25 per or more is acquired and who after such acquisition is left with 1.5 acres or less of irrigated land or 3 acres or less of unirrigated land.

(*ii*) Whose house and accompanying structures are acquired and thereby he/she is rendered houseless.

(*iii*) Any landless and houseless person who is dependent for his/her livelihood, by manual labour, on such agricultural land which is programmed to be submerged permanently or temporarily by the project and who will be deprived of his/her livelihood on account of acquisition of the said land.

Any person who is in possession of unobjectionable encroached land will also be included in category (*i*) and (*ii*) above.

Thus, for the limited purpose of rehabilitation, both the landed and landless will be treated on the same footing.

(*e*) *"Displaced Family"*

(*i*) A family in relation to a displaced person shall be a nuclear family (not joint) and shall include such person and his or her spouse, minor sons, unmarried daughters only.

(*ii*) Every major son (18 years and above in age) whether married or not will be treated as a separate family.

(*iii*) Physically infirm and mentally retarded persons irrespective of their age and sex will be treated as separate families.

(*iv*) Minor 'orphans' who have lost both the parents and have no body to fall back upon will be considered as separate families. More than one orphan in any particular family will be counted as one family.

(*v*) 'Divorcees' (as per court decision) with dependent child/children and divorcees without children but having no source of livelihood will be considered as separate families.

(*f*) "Fully Submerged Village" means that village or unit as recognised by Government which gets submerged completely.

(*g*) "partly submerged village" is one which gets submerged partly.

Land Acquisition

(*a*) All the lands situated below the Full Reservoir level of the Project shall be acquired completely.

(*b*) If 75 per cent or more land of a contiguous holding of any person is required to be compulsorily acquired, such person shall have the option to compel compulsory acquisition of the entire contiguous holding.

(*c*) All buildings with their appurtenant land situated between Full Reservoir level and Maximum Water Level shall also be acquired.

(*d*) After normal acquisition if it is seen that the left over land and population will not constitute a viable community and the said population desired to be shifted, their cases should be considered favourably and land acquisition made accordingly.

The present process of acquisition of land is lengthy. It should be cut short and delegation of statutory powers should be made so that most of the work ends with the L.A.Collector as far as possible.

Baseline Survey

The main objective of the resettlement of displaced persons is to restore them to their former standard of living or even better. In order to achieve this, an accourate baseline survey in the submersible area must be done and for this specialised institutions should be entrusted with the survey. The survey would include:

(a) Census of Affected Population

A census of the persons to be affected should be made well in advance before construction activities in the project commence. Date of publication of the notification under section 4(1) of the L.A.Act, 1894 should be regarded as the date of reference and survey should be made accordingly to know the actual number of individual families for evacuation and resettlement. This census should give the details of number of families, family size, age and sex of every member, their religion, educational qualification and ethnicity etc. The list should be updated as on 1st January of the year in which the actual displacement takes place. The census work should also be done with proper and advanced notice to the would be oustees and it should be ensured that no outsiders manages to get enumerated as oustee and avail of the rehabilitation benefits at the cost of genuine oustees.

(b) Inventory of Property

The property inventory to be made should contain information on extent of land (homestead, communal and agricultural), whether irrigated or unirrigated, trees standing on them, houses with number of rooms (whether pucca or thatched), farm equipments and livestock, wells, etc.

(c) Existing Standard of Livelihood

It should indicate occupation of individual member of the family (cultivation, service, artisan, wage earner, etc., with source), and average annual income.

(d) Census of Public/Community Property

The reservoir may submerge public/community property *i.e.*, roads, school buildings, dispensary, Club Houses, Temples, etc. Details of these property alongwith depreciated value should be indicated in the report.

(e) Encroachment

Details of Government land and property encroached upon (whether objectionable or unobjectionable) and cost of the property standing thereon should be furnished.

(f) ***Besides*** survey on cultural, Anthropological, educational and health status of the people coming under submerged area should also be undertaken.

Identification of New Settlement Site

As far as possible adequate and suitable land for resettlement of the displaced families should be selected in the Command Area of irrigation project concerned and the land should be free from all encumbrances so that there is no difficulty in recording the land in favour of the resettlers. The site so selected should be very close to the submergence area as far as practicable, so that there is not much of environmental differences; and those displaced persons who are left with their balance agricultural land unsubmerged can continue to cultivate them even after shifting. Further, care should be taken to resettle the tribal people in groups and for which site should be selected accordingly as far as possible very close to the area of displacement. The tribals being closely knit socially cohesive groups, it is essential that they are rehabilitated in surroundings which are similar to the area from which they are displaced. Representatives of the oustees and voluntary organisations may be involved in the process of identification of settlement sites at different places.

Host Population

Pressure on land leads to encroachment. This is intensified due to the prospects of irrigation water that would be available on completion of the water distribution system. Hence if the site selected for resettlement purposes lies in the Command area, the same is likely to be occupied by encroacher. This may create obstacle in effective resettlement of displaced persons, if not well studied and incorporated in the action plan. Detailed census of such host population is, therefore, very much necessary. Not only that, with a view to integrating the displaced and host population, resettlement plan and programme should be chalked out in such a manner that the host communities are involved from stage of planning for development of sites itself. All development related problems of host communities should be solved so that they do not feel neglected in any manner.

Development Plan for the New Settlement Site

Resettlement planning should begin immediately after the basic baseline survey is completed. Planning should be made for each displaced/affected village and resettlement site and should have a distinct, time frame for all activities. The content and level of details of any resettlement plan will vary depending on the magnitude of resettlement activities. But the plan should normally include the following:-

(*a*) Valuation and compensation for lost assets.

(*b*) Land acquisition.

(*c*) Environmental protection and management.

Normally very little attention is paid to this aspect. With the influx of resettlers to the new site, environmental issues like deforestation, overgrazing, soil erosion, health issues are likely to become serious. Hence the comprehensive R and R plan should include appropriate mitigating measures. In case of difficulties, alternative relocation sites should be identified.

(*d*) Shelter, infrastructure and social services.

(*e*) Training and employment.

(*f*) Community participation and integration with host population.

(*g*) Organisational responsibilities.

(*h*) Implementation schedule, monitoring and evaluation.

Most often our planners forget to analyse the local legal framework relevant to resettlement operations. Laws of the land relevant to resettlement activities (Land Settlement Act/ Compensation rules, etc) should be analysed thoroughly and made clear to the executing authorities.

Most important aspect in planning the resettlement is the identification of possible relocation sites.

(i) Layout Plan

The resettlement colonies should not be large aı un-weildy. An ideal colony may consist of one hundred households. Accordingly detailed engineering lay out plans should be prepared to set up colonies to accommodate all the displaced families.

The colony should be designed to accommodate the homestead plots. The plots would have space for constructing houses to suit the way of life of the settlers and additional space for poultry, dairy animals, kitchen garden, fruit and flowering trees etc.

It should have also facilities to lay out the service infrastructures like electricity, water supply, drainage, roads, school buildings, club houses, temples, etc.

Besides the agricultural lands, there should also be sufficient open space to grow woodlots for firewood purposes, fodder plots for domestic animals and space for other non-farm activities. Land for community uses and for future expansion of the colony should also be set apart clearly in the layout plan.

(ii) Land Reclamation

The land earmarked for resettlement colony (both for homestead and agricultural purposes) should be reclaimed properly. There should not be any uprooted tree stumps. It should be properly levelled. Agricultural land should also be properly levelled and made really cultivable. These works should be done by project authorities, instead of leaving the same to the resettlers.

(iii) Constitution of Committees

There should be site selection committee duly constituted by the Government. The Committee may comprise the concerned Tehsildar, the Block Development Officer, the Executive Engineer of the Development Division of the Project in question, the Sarpanch and representatives of the displaced persons. Voluntary organisations should also be co-opted. The Rehabilitation Officer will be the Chairman. The proceedings of the Site Selection Committee may be ratified by the Rehabilitation Advisory Committee. The Site Selection Committee will also identify the locations for the proposed buildings, wells and tanks and other civic amenities to be provided in the colony.

Identity Cards/Disseminating Project Information

'Praman Patras' or identity cards may be issued to all displaced heads of family in token of their displacement. The card would contain the names of the head of the family, his age, father's name, village and thana from which displaced, name and age of other family members dependent upon him. There would be an attested passport size photograph too of the person issued with identity cards. The card would be delivered on the date of shifting under proper acknowledgements.

Alongwith the issue of identity cards, informations related to the project (its salient features, benefits, etc.), the salient features of the Resettlement and Rehabilitation policy and the resettlement programme in brief must be disseminated in form of booklet. Other credible and reliable means of information dissemination should also be adopted so that the displaced persons are fully aware of their fate sufficiently before the actual resettlement process begins. Further there should be information campaigns and meetings with all the project affected persons so as to enable them to be thoroughly knowledgeable about the proposed R and R activities.

Payment of Compensation Money

Due process of law *i.e.,* Land Acquisition Act, 1894 as amended form time to time has to be followed in acquiring private land and property. Prevailing price as reflected in the records of the Sub-Registrar is followed in determining the value of land. But instances are not rare when the sellers quote low price in order to avoid heavy stamp duty. In the case of S.T.,S.C. and other weaker sections of people in the community, under-rate is also offered by rich and influential persons. It causes dissatisfaction in the minds of the displaced persons. It is, therefore, necessary to fix rate of compensation money through any special mode. This may be done possibly by capitalising annual net profit from the land to be acquired. The value of land can also be determined based on its productivity during the last 10 to 15 days. This valuation may be done by the Collector of the district, without any delegation of power. Solatium, interest, etc., as admissible under rules will be paid in addition. As regards compensation for trees, it may be determined on the basis of the income derived from tree products, if any, on a regular basis and/or the market value of the timber of the standing tree.

All compensation money due, to the displaced person will be paid well before the date of his/her shifting. In order to ensure that there is no misuse or abuse of the said money, by the awardees, accounts would be opened by them in the nearest Commercial Banks. The awardees may also be persuaded to keep the amount in fixed deposits. It has been calculated that interest of 10 per cent per annum payable on fixed deposits derives income, one and a half times the price of one agricultural crop. Arrangements should also be made to the effect that the Banks will not charge any commission for encashment of the cheques and to advance loans upto 100 per cent against the fixed deposit amount for any purpose that would be certified by the Rehabilitation Officer.

As regards public property like school buildings, Club Houses, electrical installations, etc., the depreciated value of the property submerged/affected should be deposited with the Departments/Agencies owning them for substitution at appropriate place as would be determined by them. But it must be ensured that the money so deposited by the project authorities is utilised for the purpose gainfully and timely. Difierential cost, if any, should be borne by the Department/Agency concerned.

Rehabilitation Facilities

(a) Shifting

All displaced families will be provided free transport for the house building materials which they can salvage from their old houses upto the new site of settlement.

It may be noted that the shifting to the new settlement

colony sponsored by the project authorities will not commence until the infrastructure development required to be done there have been done. Nevertheless shifting has to be ensured before the impounding of the reservoir commences. The gap between the time of payment of compensation money and the actual shifting results in wasteful expenditure of the amounts and makes the settlement of the displaced persons in the new places more difficult. On the other hand during the interim period-prospective oustees continue to cultivate their land and stay in their houses. Hence physical shifting should be commensurate with the payment of compensation money or in the event of delay in shifting, the compensation money so earmarked may withheld and utilised fruitfully on the rehabilitation works during the interim period.

(b) Allotment of Land

(i) Allotment of Government Land

The main thrust of the Rehabilitation Policy should be to allot land to all the oustees irrespective of the fact that they are landowners or landless at the time of displacement. As far as practicable this policy has been adopted in all the irrigation and hydro-power projects in Orissa. But the Forest (Conservation) Act, 1980 has made it difficult to adopt this policy any longer, as the Act puts a lot of conditions for release of forest land for non-forest uses. A fresh look has, therefore, to be given regarding indiscriminate allotment of land. The following suggestions are made.

(1) In fully sumberged villages, the displaced families losing land and/or houses will be eligible for allotment of 2 acres of irrigated/irrigable land or 4 acres of unirrigated land duly reclaimed at Government cost. It should be clearly understood that there is distinction between " irrigated" and 'irrigable". Any land which is not likely to get irrigation within next three years at the time of allotment and after physical displacement commences, should not be classified as 'irrigable'. If sufficient land is available, landless and homeless agricultural labourer families may also get this benefit.

(2) In partially submerged villages, displaced families retaining houses but losing 25 per cent or more land, and thereby suffering a reduction in their holdings to three acres or less in the same or adjoining village, will be entitled to benefit.

(3) In partially submerged villages displaced families losing houses and 25 per cent or more lands and thereby suffering a reduction in their holdings to three acres or less in the same village or adjoining village will also be entitled to this benefit.

(4) In partially submerged villages, displaced families retaining houses but losing encroached lands in the same village or adjoining village so as to render them landless, will be entitled to get this benefit.

(5) The landless and homeless families in the fully/partially submerged villages who were dependent on the land acquired there, for their livelihood will also be eligible to avail of the benefit of agricultural land in the rehabilitation area.

(6) Any person who loses land through acquisition for construction of canal system will get an equal area of land if unirrigated or half the area if irrigated subject to a limit of 4 acres of unirrigated land or 2 acres of irrigated land.

Depending on availability of land, displaced persons losing whole of their land will get priority over those losing land partially and displaced persons losing homestead land will be given developed homestead plot of 20 decimals, in the colony with all civic amenities. But since in case of tribals, there is no rehabilitation more effective than providing them with land as a source of livelihood, 'land to land' principle should be very meticulously followed. This principle is also backed by the ILO convention 107, to which India is a signatory.

A person losing his/her only house due to acquisition for canal construction will be given an equal area of house site subject to maximum of 20 decimals.

(7) Any displaced person in occupation of land and house which is termed as "unobjectionable encroachment", will be eligible to get land for agriculture and house site for residence to the extent of encroachment subject to a maximum of 2 acres of irrigable land/4 acres of unirrigable land and 20 decimals of homestead land; provided that he/she does not have any land elsewhere to fall back upon.

Complete details of the lands available with the Government for allotment to the displaced persons should be maintained with the Rehabilitation Officer in his office and before allotment, the allottees should be shown these lands to express their view point. The strategy should be to relocate particular communities as such at the new site, instead of relocating the displaced persons as individuals. Such individual relocation is likely to have Ka "Scatter effect" and the displaced persons who have known one another and can interact easily with one another should be shifted together from. the displaced area to the new site of resettlement. This will help them in adjusting to the new environment easily and quickly.

(ii) Land for Purchase

Sales transaction of land takes place almost everywhere. Farmers who own large areas in the proposed Command Area also may sell a part of it either because of their inability to maintain large areas under irrigation or to get capital for developing the land at hand. In case the Government are not in a position to allot any agricultural, homestead land or the Government lands offered to the oustee are not acceptable to him on account of valid grounds and he is also not in a position to suggest any alternative land that may be allotted to him after examining the details in the Office of the Rehabilitation Officer, the Government would help these oustees in purchase of private land from the amount of compensation payable to them. For that purpose a Committee known as "Purchase Committee' may be constituted under the Chairmanship of the Rehabilitation Officer. The 'Site Selection Committee' to select site for establishment of settlement colonies may also function as the 'Purchase Committee'.

The amount in excess of the amount of the compensation that the displaced person may have to pay towards the cost of the private land will be met from out of the compensation money and rehabilitation cash assistance; and if the amount is still in excess, will be treated as 'ex-gratia' to be paid by Government.

For effective implementation of the land purchase scheme, there should be some legal restrictions imposed on indiscriminate transfer of land in the Command Area. For example, no sales transactions should be finalised without the approval of the Purchase Committee. Ceiling surplus land should also be identified and allotted to the oustees.

Further there should be full exemption of registration fees and stamp duty to the oustee families who purchase land within three years from the date of receipt of compensation money. For the land allotted by project authorities no registration fee should be charged.

(iii) Irrigation

Land allotted in the irrigation command of the project will get irrigation. But agricultural land not likely to get irrigation from the Project should also be provided with irrigation; either through lift points or through minor irrigation at the Government cost.

There may be several families who are left with some cultivable lnd after acquisition of remaining land for the project. Their sacrifice in this regard should not be ignored. The land retained by them close to the reservoir, may also be provided with left irrigation facilities.

(iv) Allotment of Non-agricultural Land

Not that all the resettlers will depend on cultivation. Not also that land for cultivation will be available to all in the colony. Hence opportunities should be explored to allot waste land/degraded forest land for raising commercial forest plantations on lease basis. The extent may be five acres in the maximum. In lieu of reclamation of agricultural land for which Government pays the cost, money for development of the land for plantation may be given. The tribals especially, whose living is mainly based on forest should be enlisted to take to forestry.

(c) House Construction

It is always desirable that the oustees build their own houses, so that there is a sense of belongingness in them. A reasonable amount of money as house building assistance may be given to the oustees whether or not they resettle in project sponsored colonies. Self-reliant low cost, employment generating and resource conserving building practices should also be encouraged so that the resettlers can utilise efficiently the locally available resources like soil, timber, bamboo, lime and other farm fibres. This assistance in shape of money may be given in suitable instalment subject to verification of proper utilisation of previous instalments. The project authorities may also prescribe specifications of house to be constructed by oustees. However, such specifications should not stand as hindrance for the oustees to get instalments released. House building forest materials should be made available by the State Forest Deptt., from the nearest forest on payment of a single royalty at tenant's rate to all the oustees irrespective of the fact whether they resettle in project sponsored colonies or not. The oustees requiring cement and steel etc. For construction of houses may also be given permits for purchase from authorised dealers.

(d) Civic Amenities

The following civic amenities should be provided in the sponsored resettlement colonies:

(1) One two roomed primary school for every 100 families or less. There will be a drinking water well too in the school campus.

(2) Drinking water well with trough at the rate of one for every 20 families.

(3) One village pond for each colony irrespective of number of families resettled there.

(4) One Community Centre in the colony irrespective of the number of families. There will be a drinking water well also near it.

(5) One dispensary for every 500 families.

(6) Each colony should be linked to main road by all weather roads.

(7) Each and every colony should be electrified.

(8) Amenities like school for higher education, Veterinary Dispensary, Panchayat Ghar, etc., should also be provided by the respective Administrative Departments duly being supplemented by the project.

According to the father of the nation Mahatma Gandhi, on villages; "the unit of society should be a village, or call it a manageable small group of people who would, in the ideal, be self-sufficient (in the matter of their vital requirements) as a unit and bound together in bonds of mutual co-operatives and interdependence. "Hence the colonies set up by the project authorities must contain all civic facilities necessary for a decent community living. The Government should also pay equal attention for the development of the area near about the settlement colonies so as to create a healthy guests (resettlers) and hosts relationsip. Most often the host community get sore of the development infrastructures created in settlement colonies because due attention is not paid for such activities in their villages.

In case of oustees opting to make their own rehabilitation arrangements and resettling in groups or clusters of ten and more number of families; minimum civic amenities of drinking water wells/tube wells, tanks, etc., should be provided at Government cost.

(e) Subsistence Allowance

It is very natural that the displaced persons will have to

wait until new lands are cultivated and first crop is harvasted or they get steady source of livelihood. In order to compensate for the wages lost at the old site and to meet the daily necessities, each oustee family should be paid a subsistence allowance as would be prescribed by the Government per month, for a period of one year to be counted from the actual date of shifting. This will be subject to the condition that the displaced family shifts from the area in question as per the programme framed by the project authorities.

(f) Alternative Strategies

India is primarily an agrarian country. Hence land represents the means of livelihood of the majority. Therefore, it is natural that the oustees losing land due to submergence will demand land as replacement. But it may be known that almost all productive lands have already been put to the plough. Forest land that can be converted to agricultural land is denied for resettlement purpose under Forest Conservation Act; unless and until equivalent non-forest land is made available for compensatory afforestation. Hence alternatives are to be found out. Of course, it should be mandatory to allot homestead land to each and every oustee for his shelter. The Resettlment and Rehabilitation programme should be seen as an integrated process which not only includes provision of housing and basic civic facilities, but also measures to strengthen Community organisations, promote to restore income generating activities, provide vocational training and social assistance services aimed at ensuring the long term viability of the communities in the new and changed environment. The viable alternatives to land for cultivation could be:-

(i) Tree Culture

It has been discussed elsewhere that there is rapid loss of forest cover in the country. No forest law, however stringent it may be, could possibly check it. Experience teaches that rehabilitation programmes in many projects have not been able to restore the community life and provide adequate means of livelihood to the displaced persons. One major factor responsible for this is that, the tribals and other weaker sections of the affected people who, before displacement used to meet many of their needs from the forest, do not adjust in the non-forest environment on receipt of cultivable land/cash compensation. Their previous subsistence economy gave them no scope also to take to irrigated agriculture in the new place. This abrupt change leads to mental uneasiness. In this background, it is necessary that tribals and those people who were considerably dependent on forest economy, and those who are landless, should be rehabilitated with forest bias so that they can enjoy exclusive right of collection of the minor forest produce. This is how "tree culture" can also develop. Some call it "bio-mass" culture. It can be developed both in the periphery of the reservoir and in the Command Area; and also elsewhere in Government waste land, degraded forest areas, etc.

(1) Periphery Tree Culture

It is argued that cultivation of the periphery land will increase silt load in the reservoir and hence should not be encouraged. There are some who do not subscribe to this theory. They advocate for leasing out periphery land where it is comparatively flat. Cultivators, owning land in the reservoir periphery and catchment area can be motivated to grow forest. Government waste land and degraded forest land can be leased out to the oustees for such purpose.

(2) Tree Culture in Command Area and Elsewhere

The land allotted to the resettlers in colonies are not very fertile. Further they all are not irrigable. Resettlers who do not have aptitude for agriculture can be motivated to take up forestry including irrigated forestry. Borrow land along both the sides of canals can also be leased out in any manner. Clauses will be provided in the lease deed to the effect that the canal authorities will have freedom to use earth from the borrow area and will terminate the lease in the event of necessity, after giving due notice in advance.

The activities relating to the social forestry scheme of the State Government are mostly confined to avenue plantation.

Individual pursuit has been lacking so far. It is necessary that Government gives inducements to raise trees which can yield more income than conventional agriculture. Tea and coffee plantation can be done on co-operative basis. Mulberry plantation both under irrigated condition and rainfed condition is equally profitable. More and more efforts should be made to raise trees yielding later (rubber), trees yielding lac, trees yielding resins, waxes and lubricants, tree yielding pesticides ("neem" and "karanj") and trees yielding non-edible oils. K.R.Datya of Centre for Applied Systems Analysis in Development, Bombay in his "Alternative strategies for rehabilitation of tribals in reservoir fringes—A case study for Ukai reservoir" *writes as under:*

> "Income from non-edible oils can be far in excess of incomes from fuel wood and timber. A seed yield of 0.5 kg per year, with a plantation density of 2500 trees per hectare, provides 1250 kg of seed which can be processed to provide 500 kg of oil, if sold at Rs.20 kg. This gives an income of Rs.10,000 per hectare. Tree survival and early maturity can be ensured. Through horticultural means these could be brought down to 3 to 5 years. Non-edible oils provided raw materials for such downstream industries like soap manufacture, production of chemical intermediates and chemical conversion to edible products."

The cropping pattern to be modelled should contain programme for growing cereals, cash crops and trees, so that the resettler can become self-sufficient in food, fodder and fuel wood/timber yields.

(ii) Cottage Industries

Cottage industries wherever possible may be set up to provide livelihood for the resettlers.

Agro based rural industries have begun to play a vital role in the development of rural areas by providing income generating employment. "Sisal" (Agava Veracruz) plant is a very potential raw material for agro based industries. Large scale sisal plantation in the catchment area as well as in the non-agricultural land will not only help checking soil erosion but will also provide raw materials for three important categories of industries (1) the sisal leaves which yields long fibres can be processed to manufacture ropes,;(2) the waste bagasse after being separated from the sisal leaves can be crushed into juice to prepare "hecogenin" and (3) the sisal waste after decortication to prepare ropes, also contains fibrous matter which can be used to manufacture paper on a small scale. The Karnataka State Council for Science and Technology have developed and propagated such "sisal utilisation" projects in Karnataka. (*Source*-Annual Report 1985-86 of Karnataka State Council for Science and Technology).

(*iii*) Granting milch cows and buffaloes, goats and sheep, poultry, etc., at subsidised rates will also help the resettlers in earning their livelihood.

(*iv*) Floating net cage pisciculture in reservoir areas is a recent development which should be practised and encouraged. Fising rights should also be reserved in the reservoir for the displaced persons who may organise themselves into co-operatives. In this regard preference may be given to tribals and weaker section of the displaced persons.

(*v*) *Vocational Training:* The resettlers who are not farmers may wish to continue in thier traditional occupations as blacksmiths, tailors, barbers, etc. There may be also others who do not belong to these trades, but want to take them up as their occupation. Hence it is necessary to provide free training facilities in unskilled and sem-iskilled jobs by setting up schools at convenient places in the colonies. In the alternative, seats for the eligible resettlers may be kept reserved in nearby industrial training institutes. Of course, there will be no commitment that such training opportunities afforded will bind the Government/Project authorities to provide employment. However, jobs as and when available in the project should be offered to those trained displaced persons in perference to others. Further it should be the endeavour of Goverenment to provide resources for self-employment schemes.

(*vi*) At the construction stage, the project affords innumerable employment opportunities. Enumerated displaced persons should be given first opportunity in this regard. Besides one eligible member of the displaced family should be given priority in employment either in any Government Organisation or in any public undertaking. For this, necessary statutory provision has to be made to bind the Government in general and the project in particular to reserve seats for the displaced persons. Project displaced people should also be given concession in age in recruitment to different services. In the event of priority employment given, excepting homestead land, no cultivable land for agriculture should be given to that family. The following priority may be followed in providing employment.

(1) Oustees whose lands including houses are fully submerged. Among such oustees, S.T and S.C. oustees will be given top priority.

(2) Oustees whose lands including houses are partially submerged.

(3) Persons whose lands are acquired for purposes other than submergence.

(4) Allotment of Shops / Houses.

At project cost, buildings for locating shops for different commodities may be constructed on Government land close to the project and allotted to those oustees who do not have aptitude for cultivation or employment, on long term lease basis. Hence agricultural land will not be given to them. They will be given homestead land only. Finance if required for running the shops should also be arranged by Government for these oustees. Built up shops should also be allotted to the resettlers in the colonies after assessing the requirement. Of course, the resettlers who have been allotted agricultural land will not be debarred from getting this benefit. But their priority will come next to those settlers who have not been allotted agricultural land.

Priority should also be accorded to the displaced persons in allotment of house under Housing Schemes.

(g) Development Programmes

In order to assist the resettlers to regain or improve the previous standard of living, secondary employment sources should constitute a major thrust in all resettlement programmes. After the infrastructural and operational linkages are organised during the stage of development of resettlement sites itself, the arriving resettlers should be covered under the following financial and technical assistance programmes. The Resettlement and Rehabilitation Officer should be made responsible for this.

(*i*) Land development and levelling.

(*ii*) Distribution of modern agricultural implements.

(*iii*) Distribution of plant protection equipments.

(*iv*) Distribution of bullock carts/trollies, rickshaws.

(*v*) Distribution of milch cow and buffaloes.

(*vi*) Distribution of sheep and goats.

(*vii*) Distribution of biogas plants/solar cookers.

(*viii*) Providing scope for pisciculture, poultry.

(*ix*) Incentives for setting up Agro-based industries.

(*x*) Incentives to set up shops, including tailoring units, blacksmith and Carpentry Units etc.

Option will be provided to each oustee family to select any one productive scheme of its choice from the above programmes within financial limit as would be fixed by the Government. Subsidy as would be prescribed by the Government should also be given for items No.(*ii*) to (*x*) in kind. For the tribal oustees and Scheduled Caste oustees and those are not able to repay the loan because of their weak economic condition, higher percentage of subsidy may be allowed.

(h) Model Towns

Model small towns with all infrastructures may be set up in central places so that the non-agriculturists can also get opportunities to earn their livelihood by engaging themselves in different non-agricultural vocations.

Rehabilitation Cash Grant

Although land for land should be the guiding principle of resettlement and rehabilitation of oustees, there may be many of them who would like to be on their own and instead of moving to the project sponsored colonies would like to go to places of their choice. Restraining them to do so, may amount to encroaching upon their fundamental rights. But experiences have already taught us that the allurement of cash payment in lieu of allotment of land has stood on the way of successful and planned implementation of rehabilitation schemes in project sponsored colonies. Secondly, for want of adequate follow up measures to ensure that these oustees really make their own rehabilitation arrangements, more and more socio-economic disorder is likely to develop. Hence it should be the endeavour of the Resettlement and Rehabilitation Officer to persuade the oustees to avail of the benefits of colony life. In order to achieve this, the benefits in the colony should be made really alluring. Further it should be made clear to the oustees that declining to avail of colony life and allotment of land by the Government will deprive them to get government land elsewhere for agriculture, as landless people. Necessarily this would be recorded too in the identity cards issued to the oustees. However, homestead land should be given to one and all oustees, irrespective of the fact that they reside in project sponsored colonies or outside.

Those oustees who will not opt for colony life and will be on their own, will also be given rehabilitation cash grant as would be prescribed by the State Government. This will be over and above the compensation money towards acquisition of land and property. part payment in cash may be made to those resettlers who are allotted part land, but this should not be done indiscriminately. In Upper Kolab Project very small holdings have been allowed which are not ecomically viable. Hence while allotting part land, viability aspect should be seen. But in case of tribals, it should be the complete responsibility of the project authorities to relocate and re-settle them. In this regard, representatives of State Tribal Welfare Department should be closely associated. For that matter, a cluster apporach under the Tribal Sub-Plan can be applied too in case the number of tribal oustees is more than 25 per cent or so of the total number of oustee families. Such arrangement will help in payment of contiuned attention to the displaced tribals by the State Tribal Welfare Department as well as by the Ministry of Welfare, Government of India.

Detailed accounts of all the oustees receiving cash grant should kept and their mode of effective rehabilitation should be monitored as far as practicable. Besides, issue of identity cards, to the oustees, the list of oustees receiving only cash grant should be transmitted to all the district Collectors in the State for reference and record.

Review/Monitoring

The Resettlement and Rehabilitation Officer will review the progress of evacuation of displaced families and their effective resettlement regularly every month. He will ensure the following:-

(*a*) Shifting has been done as per schedule and the oustees have got their dues in time.

(*b*) Allotment of land and provision of civic amenities in the colonies have been made as per norm and as per schedule.

(*c*) "Pattas" are distributed in time and encroachment, if any, is removed.

(*d*) The assets created in the colonies are transferred to the respective administrative departments well in time so that because of delay there is no difficulty in maintaining them. This is very much necessary because after completion of the project construction period, the civic amenities created remain uncared for. Lack of timley maintenance and repair results in either complete damage or heavy expenditure on repair.

(*e*) Development programmes as contemplated are in execution to the full advantage of the resettlers.

Besides the Rehabilitation Advisory Committee as discussed, there should be also a High Power Committee at the State level under the Chairmanship of Chief Secretary to review and monitor the progress of activities.

Responsibility of the R and R organisation in the matter of resettlement and overall development of the resettlers is a time bound activity. It ceases almost with the completion of the project activities. Hence it is necessary that this responsibility is taken over by the respective line department of Government. Most often it is seen that the shift of responsibility from the Project authorities to the line departments is very slow. Hence it is necessary to associate the line departments in resettlement programmes from the very beginning and the progress of shift of responsibility should be monitored regularly.

Another important aspect to monitor is to see that the resettled population are not forced out by real estate speculation once infrastructure and other basic services are completed in the resettlement sites.

Organisation

(a) Government Organisation

(*i*) At the project level there should be one independent officer each for Land Acquisition and Resettlement and Rehabilitation, alongwith supporting staff of different disciplines. Of them the important disciplines are Agriculture and Co-operation, Soil Conservation, Small Scale Industries, Animal Husbandry and Pisciculture. There will be a distinct development/construction wing also under it to execute programmes. The Rehabilitation Officer may be a senior Administrative Officer.

He is the Chief Co-ordinating Officer between all the line Departments in the matter of planning and programming of resettlement programmes and implementing them too. Only officers and staff who are genuinely interested in the human problems of the oustees should be selected and kept in charge of the implementation of the programme. Incentives should be given to the best workers so as to attract talent and enthusiasm.

(*ii*) There will be a Rehabilitation Advisory Committee as is existing today with slight modifications to include representatives of different disciplines and voluntary organisations related to human development.

The main functions of this Committee *inter alia* will be to examine the rehabilitation schemes prepared for the oustees, to review their implementation schedule and results, recommend improvements, if any, etc. This Committee will also ensure that the environment lost, if any, due to construction of the project and setting of settlement colonies is appropriately compensated and there are sufficient safeguards against deterioration of physical as well as social environment.

(*iii*) At the Government level there should be a separate Directorate of Land Acquisition/Resettlement and Rehabilitation to supervise the works in respect of all development projects.

(*iv*) The responsibility of taking final decision of any matter should be vested in the Spl.L.A.O/Rehabilitation Officer as far as practicable. Limited issues requiring policy decisions only may be referred to higher up.

(b) Community Organisations

In order to organise the resettlers to educate and motivate them to provide group support and to help them to choose any vocation which can be used by them productively, colony-wise organisations should be constituted. They will be provided with opportunities to acquire skills in community organisation, basic knowledge in application of modern agriculture implements, pesticides and fertilisers, animals care, etc. These Committees will also endeavour that the resettlers in the colonies are accepted by the host population close by, as their partners in every walk of life.

(c) Voluntary Organisation

Non-Government voluntary organisations should be involved totally from the start of shifting till effective re-settlement and rehabilitation of the oustees. Women's participation wherever possible should be encouraged. It is women who are to a very great extent responsible in making the natural resources productive and thereby enriching their families in particular and society in general. Instead of involving any and every organisation, those organisations having experience in rural and community development works should be associated. Political consideration, if any, should be avoided. It should also be endeavoured that involvement of such organisations if turns into conflicts between the resettlers and project authorities should be avoided. For that matter the involvement of these organisations should be well defined.

Evaluation

Resettlement is a great social and economic process involving responsibility. It needs umost care and humanitarian touch. There should be prepared a chronology of all the Resettlement and Rehabilitation activities too of a project so that a perspective of the activities can be had at a glance. An independent evaluation work should be conducted by an organisation not associated with the project in any manner, to obtain an impartial view of the problems encountered and the ways of resolving them. Acceptable recommendations made by this body should be given due consideration for implementation.

Budget and Time Table

Expenditure for various components of land acquisition, resettlement and rehabilitation activities programmed should be calculated properly and tabulated. To add to it charts and diagrams should be drawn to indicate the implementation of development activities. It is to be ensured that all infrastructure development activities for resettlement of the oustees, were complete before physical transfer from the submergence area commences.

The attitude of treating the oustees as mere objects to be quickly displaced from the submersible area should be changed and they should be treated as part of a human problem and should be dealt with by compassion and concern. Primary importance should be placed on taking care of the needs of the oustees before even a decision for the implementation of the project is finalised. The objective is that the reservoir is for the benefit of everybody, not a curse.

An attempt has been made here, to go into the entire gamut of rehabilitation policy issues relating to oustees of water resources projects and to suggest some modifications and improvements on the prevailing situation. Instead of limiting their application to the situations prevailing in Orissa, they can be attempted elsewhere too. The following suggestions may also be considered.

(*a*) At the national level, Government of India should formulate policy parameters for resettlement of oustees of all the water resources projects in the country.

(*b*) In respect of inter-State rivers/projects inter-State rehabilitation committees may be constituted for effective inter-State co-ordination.

(*c*) At the State level, policy should be guided by a regional/local approach peculiar to the area. Instead of having a micro programme, applicable to all projects and all families, it should be location specific and group specific so that individual needs of the groups of people affected due to the construction of the project can be dealt with appropriately. In other words the rehabilitation scheme should be a comprehensive package recognising all the variables concerned and also flexible enough to incorporate modifications at a subsequent stage.

Annexures

Annexure-I
Rates of Payment for Compensation of Houses, Wells, Tanks and other Structures.

S.No.	*Classification*	*Rate of plinth & flooring per sq.m / sq.ft.*		
		Total rate (room & verandah).	*Rate per room*	*Rate per verandah.*
1	*2*	*3*	*4*	*5*
1.	Pucca naria tile roof, cement mortar C.B.brick wall, cement floor.	Rs.167.91	-	-
2.	Kachha thatched roof, mud wall house.	Rs.39.38	Rs.54.96	Rs.23.48
3.	Kachha naria tile roof, mud wall house.	Rs.47.15	Rs.62.30	Rs.29.10
4.	Kachha-pucca naria tile roof with C.B.brick & mud mortar house.	Rs.75.16	Rs.101.05	Rs.26.90
5.	kachha-pucca naria tile roof with C.B.brick mud mortar wall house.	Rs.79.39	Rs.107.00	Rs.31.25
6.	Kachha thatched roof with jhati-mati wal.	Rs.29.72	Rs.49.87	Rs.19.43

(contd.)

Annexure-I (Contd.)

1	*2*	*3*	*4*	*5*
7.	Kachha thatched roof without wall.	Rs.26.40	-	-
8.	Kachha naria tile roof without wall.	Rs.31.70	-	-
9.	Kachha thatched roof with planked walls.	Rs.28.22	-	-
10.	Kachha naria tile roof with planked walls.	Rs.32.50	-	-
11.	Pucca slab roof or A.C. sheet roof building with brick masonry wall.	Rs.322.80	-	-
12.	Brick or stone packing mud mortar compound wall.	-	Rs.24.78/sqm	-
13.	Pucca wall			
	(a) Stone or brick cement mortar pucca well (inside volume of the well).	-	Rs.167.75/cum	-
	(b) Kachha pucca well or brick packing kachha-pucca well. (inside volume of the well)	-	Rs.51.34/cum	-
	(c) Kachha pucca well with bullah. (inside volume of the well)	-	Rs.65.00/cum	-

(contd.)

Annexure-I (Contd.)

1	2	3	4	5
	(d) Kachha well only earth work (inside volume of the well).	-	Rs.5.42/sqm	
Notes:-				
	1(a) Support (i) Plank make	-	Rs.26.90/sqm	-
	(ii) Bamboo make	-	Rs.13.45/sqm	-
	(b) Platform for food grains-		Rs.26.30/sqm	-
	2. Floor For items 2 to item 10 extra payment will be made on actual measurement for pucca flooring.	-	@ Rs.12.85/sqm	-
	3. For plastering(pucca) for item No.4&5, extra payment.	-	@ Rs.1.77/sqm	-
	4(a) For items 3,5,8 & 10, if cement tile is used in place of naria tile, extra payment will be made.	-	@ Rs.5.45/sqm	-
	(b) For items 3.5 & 10, extra payment for mangalore tile in place of noria.	-	@ Rs.0.50/sqm	-
	5. Extra payment for naria tile in place of thatched roof in item No.6.	-	@ Rs.2.25/sqm	-
	6. Mixed walls of stone earth			

(contd.)

Annexure-I (Contd.)

1	2	3	4	5
	jhati-mati etc., payment will be made taking into consideration the higher quantity of concerned item.			
	7. For slate roof or BCI roof, payment will be made under item No.11.			
	8. For double storied buildings, payment will be made under item No.11.			
	9. Temple buildings-payment will be made as per actual measurement.			
	10. Water outlet walls—payment will be made as per actual measurement.			
	11. Depreciation cost-			
	(a) For kachha house-20% deduction			
	(b) For pucca-kachha house-25% deduction.			
	(c) For pucca house-2.5% annual deduction limited to 30 per cent.			
	12. As the acquisition will be done compulsorily, 15% excess payment will be made over and above the rates.			
	Rates paid per square metre of plinth area of different types of houses other than prescribed in the norms of Rengali Project.			
	1. Thatched roof house with R.R. stone masonry in clay in foundation and plinth mud wall with bamboo jafry ceiling and all the wood work in nonsal wood.			Rs.34.75/sq.m
	2. Thatched roof house with puddle clay in the foundation & plinth wattle and mud wall including 12 mm. thick mud plaster on both side wall, wood work in seasoned and sawn sal wood.			Rs.44.10/sq.m.
	3. 3"thatched roof on nonsal bullah with bamboo framing fixed to ground plinth of 0.5 m high or rubble packing-no side wall (cow-shed)			Rs.12.00/sq.m
	4. G.C.I.sheet roof house over nonsal wood frame with mud wall and bamboo ceiling.			Rs.84.00/sq.m.

Annexure-II
Details of Land (Private Land) Acquired and Land Acquisition Charges Paid (Original)

Sl No.	*Name of the village.*	*Area acquired in acres*				*Compensation paid (in Rs.)*		*Total expenditure (in Rs.)*
		Agricultural.	*Homestead.*	*Other communal.*	*Total*	*Land / Trees*	*Houses*	
1	2	3	4	5	6	7	8	9
1st Phase								
1.	Puki	278.58	5.34	0.66	284.58	2,98,818.40	1,31,528.00	4,30,346.40
2.	Semla	354.58	1.35	5.18	360.64	3,69,469.83	25,640.32	3,95,110.15
3.	Koranga	756.79	32.12	25.23	814.14	8,63,671.75	4,39,827.00	13,03,498.75
4.	Kechela	836.68	20.00	46.41	903.09	14,42,716.92	2,44,741.12	16,87,458.04
5.	Thalaput	311.94	2.50	10.43	324.87	3,14,226.78	1,27,296.41	4,41,523.19
6.	Duruguda	110.77	—	—	110.77	1,21,699.95	—	1,21,699.95
7.	Putukoranga	139.69	6.79	7.25	153.73	93,729.01	58,246.05	1,51,975.06
8.	Sirisi	391.79	14.10	10.63	416.52	3,37,924.14	1,69,968.62	5,07,892.76
9.	Bhitterchenda	130.55	—	—	130.55	61,874.21	—	61,874.21
10.	Bhoiguda	207.82	6.71	0.42	214.95	2,24,825.00	67,601.54	2,92.426.54
11.	Kalchuru	412.68	4.98	14.31	431.97	4,34,382.57	1,53,658.66	6,88,041.23
12.	Suka	537.85	—	1.03	538.88	3,40,779.69	—	3,40,779.69
13.	Pondi	1377.77	6.43	14.47	1396.67	23,23,411.89	5,73,384.48	28,96,796.37
14.	Thataguda	122.89	3.61	0.28	126.78	1,59,677.62	62,180.00	2,21,857.62
15.	Paidaput	512.23	8.88	13.40	534.51	3,93,310.59	94,936.00	4,88,246.59
16.	Chandalaguda	241.08	4.20	30.76	276.04	3,21,514.68	31,697.27	3,53,211.95
17.	Khalap	742.96	16.29	135.44	894.69	12,46,476.76	3,45,016.88	15,91,493.64
	Total	7466.18	133.30	315.90	7915.38	93,48,509.79	25,25,722.35	1,18,74,232.14

(Contd.)

Annexure-II (Contd.)

1	*2*	*3*	*4*	*5*	*6*	*7*	*8*	*9*
2nd Phase								
1.	Landiguda	0.06	12.21	5.40	17.67	685.00	1,95,410.50	1,96,095.50
2.	Disarikharaguda	283.45	9.68	0.74	293.87	7,75,400.20	1,42,081.20	9,17,481.40
3.	Tiadiput	727.11	19.12	4.81	751.04	18,95,745.05	1,42,557.25	20,38,302.30
4.	Gopalput	624.76	15.27	1.50	641.53	14,50,692.68	2,99,687.22	17,50,370.90
5.	Machaput	200.54	8.12	-	208.66	4,52,478.34	89,415.00	5,41,893.34
6.	Balada	527.31	14.72	14.44	556.47	14,10,645.83	1,92,316.77	16,02,962.60
7.	Sindiput(Nandapur)	97.21	3.06	9.01	109.28	1,58,303.03	52,224.00	2,10,527.28
8.	Malipeta	150.71	3.44	1.93	156.08	3,36,233.03	2,32,896.60	5,69,129.53
9.	Dhadi	343.12	11.86	2.17	357.15	4,36,454.64	2,71,362.00	7,07,817.24
10.	Ghataguda	102.15	-	3.88	106.03	10,35,641.74	51,365.60	10,87,007.34
11.	Mohanpada	211.41	-	1.34	212.75	12,84,602.30	99,509.10	13,84,111.40
12.	Gatanput	266.04	-	24.69	290.73	10,22,437.32	11,192.68	10,33,630.00
13.	Dumuriguda	57.61	3.00	-	60.61	93.916.31	87,455.17	1,81,381.48
14.	Dumuriput (Semiliguda)	171.67	-	11.44	183.11	5,77,387.28	-	5,77,387.26
15.	Pakhanput	109.22	3.78	-	113.00	85,849.96	-	85,849.96
16.	Semiliguda	59.36	-	2.50	61.86	68,318.48	-	68,318.48
17.	Chenda	56.45	-	0.98	57.43	54,778.61	-	54,778.61
18.	Parajamundar	132.19	-	2.60	134.79	3.30,325.95	27,098.30	3,57,424.25
19.	Teraguda	141.08	-	0.08	141.16	1,15,281.82	-	1,15,281.82
20.	Chalanput	70.26	-	1.40	71.66	95,672.55	-	95,672.55
21.	Tikarapada	267.92	-	-	267.92	5,22,418.52	-	5,22,418.52
22.	Ruseiput	259.94	-	4.38	264.32	3,42,649.50	-	3,42,649.60
23.	Rajubadei	235.25	3.75	1.37	237.37	5,10,132.93	30,717.70	5,40,850.63
24.	Patraput	157.69	-	0.22	157.91	1,64,284.90	-	1,64,284.90

(Contd.)

Annexure-II (Contd.)

1	2	3	4	5	6	7	8	9
25.	Haladiput	119.72	7.00	2.43	129.15	1,23,321.50	2,05,165.50	3,28,487.00
26.	Khamari	221.48	4.97	4.75	231.20	3,66,182.16	77,314.60	4,43,496.66
27.	Podapalama	252.53	–	–	252.53	5,14,714.34	–	5,14,714.34
28.	Gulel	90.22	–	–	90.22	1,45,563.50	–	1,45,563.50
29.	Dumuriput (Nandapur)	143.56	–	0.65	144.21	2,91,190.35	–	2,91,190.35
30.	Mundaguda	705.95	0.37	0.80	707.12	9,43,063.88	8,386.95	9,51,450.83
31.	Kangra	328.33	–	1.66	329.99	11,19,380.85	–	11,19,380.85
32.	Padagada	187.51	–	1.78	189.29	2,70,991.19	–	2,70,991.19
33.	Bankeiput	90.19	–	4.15	94.34	93,966.13	–	93,966.13
34.	Kichaiput	120.55	–	2.17	122.72	1,42,039.24	–	1,42,039.24
35.	Kumbha	300.83	–	0.33	301.16	6,43,308.25	–	6,43,308.25
36.	Kadama	238.94	–	1.41	240.35	2,11,086.66	–	2,11,086.66
37.	Tentulipada	305.80	0.52	4.38	310.70	15,16,984.46	11,605.19	16,28,669.65
38.	Kuleo	339.12	1.90	9.20	350.22	17,83,957.93	10,509.92	17,94,467.85
39.	Dangadeula	240.90	2.15	11.06	254.11	5,47,641.63	19,146.10	5,66,784.85
40.	Teraguda	127.59	0.56	0.18	128.33	3,23,680.00	30,884.88	3,54,565.18
		9062.73	125.48	139.83	9,328.04	2,22,57,408.47	22,88,299.95	2,45,45,708.42
3rd Phase								
1.	Kokiriguda	20.32	–	–	20.32	17,949.27	–	17,949.27
2.	Sapsil	0.21	–	–	0.21	75.70	–	75.70
3.	Nighamaniguda	70.97	–	0.46	71.43	1,08,107.16	–	1,08,107.16
4.	Lunguri	16.63	–	–	16.63	26,622.94	–	26,622.94
5.	Lakhaput	25.73	–	–	25.73	41,529.84	–	41,529.84
6.	Kancheiput	36.70	–	–	36.70	63,932.93	–	63,932.93
7.	Chatra	1.40	–	–	1.40	3,513.95	–	3,513.95

(Contd.)

Annexure-II (Contd.)

1	2	3	4	5	6	7	8	9
8.	Limika	38.02	-	-	38.02	57,520.67	-	57,520.67
9.	Mania	2.29	-	-	2.29	2,174.20	-	2,174.20
10.	Barangaguda	112.65	-	3.75	116.40	82,773.35	-	82,773.35
11.	Sindhiput	89.76	1.41	2.95	94.12	97,565.59	-	82,773.35
12.	Jangamput	38.01	-	-	38.01	82,884.60	23,770.80	1,21,336.39
13.	Khillaput	46.72	-	-	46.72	95,333.32	-	82,884.60
14.	Khaparaput	75.08	-	0.29	75.37	2,90,903.40	-	95,333.32
15.	Sriramput	15.68	-	-	15.68	30,623.90	-	2,90,903.40
16.	Machara	185.63	4.08	0.38	118.01	3,61,456.50	-	30,623.90
17.	Pitaguda	80.41	-	-	68.41	1,41,434.33	06,047.10	3,57,583.60
18.	Durla	24.08	-	-	24.08	21,452.90	-	1,41,434.33
19.	Pangiput	0.21	-	-	0.21	121.20	-	21,452.90
20.	Mohadaiput	93.45	-	6.53	99.98	1,52,068.37	-	121.20
21.	Phuladoba	28.97	-	-	28.97	19,416.88	-	1,52,068.37
22.	Chandalamunda	26.68	-	-	26.68	88,286.00	-	19,416.88
23.	Ranikona	25.35	-	-	25.35	48,950.45	-	88,286.00
24.	Mesingput	76.29	-	3.84	80.13	97,889.85	-	48,950.45
25.	Godiput	87.49	-	2.13	89.62	1,22,882.80	-	97,889.85
26.	Umuri	180.73	0.56	0.75	182.04	4,19,606.00	-	1,22,882.80
27.	Lanjisumu	0.80	-	-	0.80	4,922.00	-	4,19,606.00
28.	Janiguda	2.14	-	-	2.14	7,479.00	-	4,922.00
29.	Kumariput	56.76	-	1.15	57.91	2,23,905.00	-	7,479.00
30.	Umbel.	85.61	-	0.58	86.19	4,06,449.00	-	2,23,905.00
31,	Hinjeiput	0.60	-	-	0.60	1,196.00	-	4,06,449.00
32.	Badamunda	0.04	-	-	0.04	188.00	-	1.196.00
33.	Kantigarh	153.37	-	6.30	159.67	2,03,288.29	-	188.00
								2,03,288.29

(Contd.)

Annexure-II (Contd.)

1	2	3	4	5	6	7	8	9
34.	Silapondi	33.43	-	0.05	33.48	85,944.61	-	85,944.61
35.	Chagam	370.38	-	0.18	370.56	23,03,453.00	-	23,03,453.00
36.	Khillaput	37.18	-	1.81	38.99	1,06,642.00	-	1,06,642.00
37.	Sankei	178.86	-	4.98	183.84	4,42,806.00	-	4,42,806.00
38.	Gharagam	115.87	-	1.17	117.04	4,50,470.00	-	4,50,470.00
39.	Pitei	154.49	-	10.39	164.88	3,75,614.95	-	3,75,614.95
40.	Rajpalma	167.96	-	-	167.96	5,14,059.60	-	5,14,059.60
41.	Dabuguda	14.73	-	-	14.73	42.856.17	-	42,856.17
42.	Bijapur	0.65	-	-	0.65	1,548.00	-	1,548.00
43.	Machaput	5.28	-	-	5.28	24,390.00	-	24,390.00
44.	Dusura	0.13	-	-	0.13	416.00	-	416.00
45.	Badalipondi	64.16	-	-	64.16	3,22,492.00	-	3,22,492.00
46.	Charanguli	151.17	-	0.24	151.41	8,65,832.00	-	8,65,832.00
47.	Badatankua	274.09	-	0.27	274.36	14,78,948.00	-	14,78,948.00
48.	Patripondi	38.32	-	0.02	38.34	1,61,298.00	-	1,61,298.00
49.	Khalpadi	153.04	2.30	0.14	155.48	6,02,105.35	59,169.65	6,61,275.00
50.	Arna	357.20	-	7.50	364.70	13,88,050.00	-	13,88,050.00
51.	Jogiput	0.42	-	-	0.42	874.00	-	874.00
52.	Cheliamunda	332.27	-	7.66	339.93	14,49,638.00	-	14,49,638.00
53.	Kukudaput	85.14	-	0.62	85.76	4,87,366.00	-	4,87,366.00
54.	Mamel	1.51	-	-	1.51	3,251.00	-	3,251.00
55.	Gumelput	73.33	-	3.59	76.92	1,89,982.00	-	1,89,982.00
56.	Parajakhudubi	140.04	-	1.27	141.31	3,02,173.00	-	3,02,173.00
57.	Parata	2.90	-	-	2.90	13,104.00	-	13,104.00
58.	Jantaput	119.37	-	0.53	119.90	3,09,526.00	-	3,09,526.00

(Contd.)

Annexure-II (Contd.)

1	2	3	4	5	6	7	8	9
59.	Debatahanjar	14.75	-	-	14.75	29,532.00	-	29,532.00
60.	Kadrimaliput	52.05	0.80	0.34	53.19	3,12,154.00	31,348.00	3,43,502.00
61.	Tusuba	48.50	-	1.91	58.41	3,09837.00	-	3,09,837.00
62.	Kotagoudaput	0.18	-	-	0.18	237.00	-	237.00
63.	Gopalput	15.41	-	-	15.41	1,00,958.00	-	1,00,958.00
64.	Mandeiput	1.99	-	-	1.99	11,398.00	-	11,398.00
65.	Mohipalput	7.30	-	-	7.30	26,255.00	-	26,255.00
66.	Limaput	32.38	-	-	32.38	60.903.00	-	60,903.00
67.	Badapeda	184.05	-	1.60	185.65	8,41,134.00	-	8,41,134.00
68.	Sahuput	86.28	-	0.36	86.64	4,82,979.00	-	4,82,979.00
69.	Kutugam	0.52	-	-	0.52	2,887.00	-	2,887.00
70.	Gulpondi	103.62	-	0.29	103.91	4,49,935.00	-	4,49,935.00
71.	Chikapar	72.54	-	-	72.54	3,22,434.45	3,15,651.00	6,38,085.45
72.	Kaki	142.99	-	1.53	144.52	12,00,985.00	11,351.00	12,12,336.00
73.	Gagarapondi	123.78	-	1.83	125.61	12,38,256.00	-	12,38,256.00
74.	Odiyaguda	21.23	-	-	21.23	54,944.00	-	54,944.00
		5398.17	9.15	77.31	5484.63	2,07,27,171.07,	4,87,340.55	2,11,14,511.62
	Grand Total	21,927.08	267.93	533.04	22,728.85	5,22,33,089.33	53,01,362.85	6,25,34,452.18

(Contd.)

II (After Resurvey)

Sl No	Village	Area acquired in Acs.				Compensation money paid in rupees			
		Agril	Home-stead	Others	Total	Land	Trees	House	Total
1	2	3	4	5	6	7	8	9	10
1.	Khilaput	29.34	0.65	0.71	30.70	1,07,520.00	Nil	47,167.00	1,54,687.00
2.	Tentulipada	21.03	0.04	5.01	26.08	80.035.00	-	9,511.00	91,551.00
3.	Godiput	37.84	0.50	0.98	39.32	1,36,527.00	-	25,762.00	1,62,289.00
4.	Kumbha	6.20	-	0.83	7.03	38,467.00	-		38.467.00
5.	Pitei	6.26	-	1.08	7.34	22,603.00	-		22,603.00
6.	Kalchur	5.36	-	-	5.36	19,511.00	-		19,511.00
7.	Suku	13.55	-	0.85	14.40	55,168.00	-		55,168.00
8.	Pondi	13.19	-	0.13	13.32	68,543.00	-		68,543.00
9.	Limika	5.87	-		5.87	25,388.00	-		25,388.00
10.	Haladiput	4.07	-	0.16	4.23	13,576.00	-		13,576.00
11.	Rajubidei	5.46	-	0.70	6.16	18,989.00	-		18,989.00
12.	Chandalamunda	3.46	-	-	3.46	27,880.00	-		27,880.00
13.	Umuri	32.39		1.36	33.75	1,53,750.00	-		1,53,750.00
14.	Bhoiguda	6.13		0.10	6.23	20,875.00	-	-	20,875.00
15.	Bankeiput	12.00	-	2.45	14.45	53,707.00	-		53,707.00
16.	Hinjeiput	0.50	-	-	0.50	2,995.00	-	-	2,995.00
17.	Tiadiput	21.07	0.58	3.76	25.41	97,155.00	-	12,549.00	1,09,704.00
18.	Badapada	6.87	-	-	6.87	34,303.00	-	-	34,303.00

(Contd.)

II (After Resurvey) (Contd.)

1	2	3	4	5	6	7	8	9	10
19.	Umbel	11.84	-	0.29	12.13	51,663.00	-	-	51,663.00
20.	Mohadeiput	19.43	0.06	0.53	20.02	1,62,374.00	-	14,355.00	1,76,729.00
21.	Gopalaput	4.20	-	-	4.20	13,726.00	-	-	13,726.00
22.	Gopulput	2.38	-	-	2.38	30,863.00	-	-	30,863.00
23.	Disarikharaguda	1.79	-	-	1.79	5,672.00	-	-	5,672.00
24.	Kicheiput	0.45	-	-	0.45	1,338.00	-	-	1,338.00
25.	Kadam	1.51	-	-	1.51	4,625.00	-	-	4,625.00
26.	Paidaput	20.18	2.62	0.82	23.62	74,683.00	-	-	74,683.00
27.	Patraput	5.15	-	-	5.15	15.703.00	-	-	15,703.00
28.	Chenda	7.66	-	0.10	7.76	23,068.00	-	-	23,068.00
29.	Arna	25.49	-	1.57	27.06	1,83,249.00	-	-	1,83,249.00
30.	Debatahanjar	1.30	-	-	1.30	4,482.00	-	-	4,482.00
31.	Gatanput	3.42	-	0.20	3.62	11,425.00	-	-	11,425.00
32.	Dangadeula	7.83	-	1.75	9.58	31,414.00	-	-	31,414.00
33.	Kantigarh	94.00	-	11.07	105.07	5,31,836.00	-	12,469.00	5,44,305.00
34.	Kadrimaliput	1.91	-	-	1.91	12,174.00	-	-	12,174.00
35.	Parajakhudubi	23.14	-	-	23.14	1,60,271.00	-	-	1,60,271.00
36.	Gumalput	44.96	-	-	44.96	2,01,924.00	-	-	2,01,924.00
37.	Mundagarh	4.70	-	0.13	4.83	14,226.00	-	-	14,226.00
38.	Thusuba	17.90	1.75	-	19.65	82.601.00	-	2,758.00	85,359.00
39.	Charaguda	4.41	0.23	0.32	4.96	17,833.00	-	16,235.00	34,068.00
40.	Balada(Doraput)	-	8.30	-	8.30	-	-	1,90,588.00	1,90,588.00
41.	Chatra	2.14	-	-	2.14	26,804.00	-	-	26,804.00
42.	Sindhiput	0.10	-	0.05	0.15	333.00	-	-	333.00
43.	Barngaguda	3.85	-	-	3.85	12,857.00	-	-	12,857.00
44.	Dhadi	11.45	-	-	11.45	37,465.00	-	-	37,465.00

(Contd.)

II (After Resurvey) (Contd.)

1	2	3	4	5	6	7	8	9	10
45.	Limaput	7.07	-	-	7.07	25,192.00	-	-	25,192.00
46.	Sahuput	22.43	0.82	-	23.25	73,599.00	-	-	73,559.00
47.	Parajamundar	3.23	-	-	3.23	14,460.00	-	-	14,460.00
48.	Pangiput	3.22	-	-	3.22	51,157.00	-	-	51,157.00
49.	Sirisi	2.63	-	-	2.63	78,730.00	-	-	78,730.00
50.	Semala	1.80	-	-	1.80	5,448.00	-	-	5,448.00
51.	Khaparaput	11.61	1.28	0.27	13.16	41,699.00	-	10,693.00	52,392.00
52.	Khilaput	3.30	0.54	0.92	4.76	26,817.00	-	46,487.00	73,304.00
53.	Phuladaba	20.86	-	-	20.86	99,687.00	-	-	99,687.00
54.	Pitaguda	16.31	-	-	16.31	1,29,806.00	-	-	1,29,806.00
55.	Dumuriput	17.54	-	5.66	23.20	79.446.00	-	-	79,446.00
56.	Kangra	16.79	-	0.85	17.64	81,082.00	-	-	81,082.00
57.	Jangamput	16.23	-	-	16.23	86.687.00	-	-	86,687.00
58.	Ruseiput	2.88	-	0.07	2.95	19,246.00	-	-	19,246.00
59.	Kumariput	8.67	-	-	8.67	42,280.00	-	-	42,280.00
60.	Chalanput	6.67	-	1.16	7.83	28,211.00	-	-	28,211.00
61.	Kokariguda	2.37	-	-	2.37	9,069.00	-	-	9,069.00
62.	Kulab	38.48	-	5.77	44.25	1,92,492.00	-	-	1,92,492.00
63.	Ghataguda	3.49	-	0.22	3.71	15,696.00	-	-	15,696.00
64.	Mania	0.85	-	-	0.85	8,655.00	-	-	8,655.00
65.	Podagarh	9.52	-	-	9.52	47,936.00	-	-	47,936.00
66.	Gulel	0.85	-	-	0.85	9,623.00	-	-	9,623.00
67.	Chagam	11.86	-	-	11.86	68,423.00	-	-	68,423.00
68.	Poadapalama	14.93	-	-	14.93	57,914.00	-	-	57,914.00
69.	Balada	21.32	-	-	21.32	67.592.00	-	-	67.592.00
70.	Pakhnaput	0.04	-	-	0.04	160.00	-	-	160.00

(Contd.)

Annexure II (After resurvey) (Contd.)

1	2	3	4	5	6	7	8	9	10
71.	Machaput	4.79	-	-	4.79	14,720.00	-	-	14,720.00
72.	Mohanpada	2.48	-	-	2.48	8,800.00	-	-	8,800.00
73.	Janiguda	6.53	-	-	6.53	56,235.00	-	-	56,235.00
74.	Charanguli	4.68	-	-	4.68	33,819.00	-	-	33,819.00
75.	Rajapalama	11.98	-	-	11.98	44,913.00	-	-	44,913.00
76.	Ranikana	14.99	-	-	14.99	1,19,517.00	-	-	1,19,517.00
77.	Silipondi	33.58	-	-	33.58	4,20,363.00	-	-	4,20,363.00
78.	Chelimunda	0.89	-	-	0.89	6,053.00	-	-	6,053.00
79.	Bijapur	5.61	-	-	5.61	83,113.00	-	-	83,113.00
80.	Jogiput	0.90	-	-	0.90	4,852.00	-	-	4,852.00
	Total	903.16	17.34	49.87	970.40	48,45,093.00	-	3,88,579.00	52,33,672.00

(Contd.)

Annexure II (Contd.)

Phase-wise Payment of Land Acquisition Compensation (ABSTRACT)

Phase	Land Acquisition				Compensation paid		
	Agril	*Homestead*	*Other communal*	*Total*	*Land and trees*	*Houses*	*Total*
1	*2*	*3*	*4*	*5*	*6*	*7*	*8*
1. First phase.	7,466.18	133.30	315.90	7,915.38	93,48,509.79	25,25,722.35	1,18,74,232.1
2. Second phase.	9,062.73	125.48	139.83	9,328.04	2,22,57,408.47	22,88,299.95	2,45,45,708.4
3. Third phase	5,398.17	9.15	77.31	5,484.63	2,06,27,171.07	4,87,340.55	2,11,14,511.6
4. Satiguda subsidi-ary dam.	90.34	18.53	48.73	157.60	1,02,010.08	1,53,254.84	2,55,264.92
5. Resurvey Grand	903.16	17.37	49.87	970.40	48,45,093.00	3,88,579.00	52,33,672.00
Grand Total	22,920.58	303.83	631.64	23,856.05	5,71,80,192.41	58,43,196.69	6,30,23,389.00

Annexure-III
List of Villages Submerged by Upper Kolab Reservoir and Families Displaced.

S. No.	Name of village	Total No. of displaced families.	Population	Manner of rehabilitation. Settled in colonies.		Received rehabilitation cash grant	
				No. of families	Area alloted in acres	No. of families	Amount (in Rs)
1	2	3	4	5	6	7	8
First phase of displacement (upto R.L.835.00M)-Monsoon of 1984							
1.	Semla	14	75	5	10.50	9	1,49,680.00
2.	Kechela	171	755	18	57.75	153	22,59.959.00
3.	Thalaput	74	375	-	-	74	10,39,960.00
4.	Koranga	322	1288	1	6.00	321	45,22,880.00
5.	Puki	113	565	-	-	113	15,86,520.00
	Total	694	3,058	24	74.25	670	95,58,999.00
Second phase of displacement (upto 850.00 M)-July & August, 1985							
1.	Paidaput	65	325	21	61.50	44	7,02,165.00
2.	Malipondi	60	345	-	-	60	8,42,400.00
3.	Putkoranga	58	322	-	-	58	8,14,320.00
4.	Sirisi	75	375	-	-	75	10,53,000.00

(Contd.)

Annexure-III (Contd.)

1	*2*	*3*	*4*	*5*	*6*	*7*	*8*
5.	Bhoiguda	42	210	-	-	42	5,89,680.00
6.	Kalchuru	64	256	-	-	64	8,58,560.00
7.	Thataguda	24	96	-	-	24	3,36,960.00
8.	Kulaba	152	608	-	-	152	21,34,080.00
9.	Chandalguda.	23	115	-	-	23	3,22,920.00
10.	Gatanpur	7	35	-	-	7	98,280.00
11.	Landiguda	78	390	5	17.50	73	10,42,437.00
12.	Dasarikha-riguda.	72	360	-	-	72	10,10,880.00
13.	Tiadiput	95	290	7	23.00	88	13,33,800.00
14.	Gopalput	141	424	11	17.50	130	19,79,640.00
15.	Machaput	56	280	-	-	56	7,86,240.00
16.	Balda (Routput)	113	384	14+1	32.22	99	15,86,520.00
17.	Sindhiput (Nandapur)	27	58	-	-	27	3,70,080.00
18.	Malipeta	108	800	-	-	108	15,16,320.00
19.	Dhadi	147	735	-	-	147	20,63,880.00
20.	Ghataguda	1	5	-	-	1	14,040.00
21.	Mohanpuda	2	9	-	-	2	20,080.00
	Total	1410	6422	58	151.72	1352	1,95,33,282.00
Third phase of displacement (upto 856.00 M) in June, 1985.							
1.	Teraguda	18	82	-	-	18	2,52,720.00
2.	Mundaguda	5	25	-	-	5	70,200.00
3.	Masiput (Dengaguda)	6	40	-	-	6	84,240.00

(Contd.)

Annexure-III (Contd.)

1	2	3	4	5	6	7	8
4.	Khamari	39	195	-	-	39	5,47,560.00
*5.	Chikapar	10	55	-	-	10	1,40,400.00
	Total	78	397	-	-	78	10,95,120.00
Displacement in June, 1987							
*1.	Masipur	9	55	-	-	9	1,26,360.00
*2.	Chikapar	155	405	-	-	155	21,76,200.00
*3.	Tentulipada	11	32	-	-	11	1,54,440.00
4.	Haldiput	29	150	2	3.00	27	4,07,160.00
5.	Dumuriguda	39	400	-	-	39	5,47,560.00
6.	Sindhiput	20	105	-	-	20	2,80,800.00
7.	Sargunaput	16	85	-	-	16	2,24,640.00
8.	Kadrimaliput	19	95	-	-	19	2,66,760.00
9.	Jaraput(Rajubedei)	18	95	-	-	18	2,52,720.00
10.	Kuturiput	24	95	3+2	7.50	21	3,36,960.00
11.	Jamuguda (Kulaba)	12	60	-	-	12	1,68,480.00
12.	Andalguda	3	15	-	-	3	42,120.00
13.	Lanjiguda (Khalapadi)	34	136	-	-	34	4,77,360.00
14.	Biriguda	28	102	28	121.50	-	60,522.00
15.	Doraput (Balda)	63	320	41	145.48	22	3,70,320.00
16.	Parojapondi	227	990	227	562.70	-	14,040.00

(Contd.)

Annexure-III (Contd.)

1	2	3	4	5	6	7	8
17.	Chagam	1	4	-	-	1	14,040.00
18.	Kaki	2	9	-	-	2	28,080.00
	Total	710	3153	301	840.18	409	68,16,342.00
Fourth phase of displacement (upto 858.00M) after June 1983.							
1.	Khilaput	21	84	21	47.83	-	-
2.	Godiput	8	40	-	-	8	1,12,320.00
3.	Mahadeiput	11	50	-	-	11	1,54,440.00
4.	Charaguda	10	55	-	-	10	1,40,400.00
5.	Tentulipada	8	30	-	-	8	1,12,320.00
6.	Pakhanput	7	29	1	1.50	6	84,240.00
7.	Khapraput	3	22	3	10.50	-	-
8.	Khilaput (Lamptaput)	1	4	1	2.00	-	-
9.	Tiadiput	26	134	15	26.12	11	1,54,440.00
	Total	95	448	41	87.95	54	7,58,160.00
Displacement scheduled in 1990 after resurvey in June 1989.							
1.	Khapraput	65	315	65	Shifted to colonies in 1991		
2.	Khilaput	39	201	39			
	Total	104	516	104			

(Contd.)

Annexure-III (Contd.)

1	2	3	4	5	6	7	8
Displacement due to creation of Satiguda Irr. Pond-in June 1982.							
1. 2.	Badliguda hamlet Dhemsaguda hamlet	80	141	-	-	80	11,23,200.00
	Total	80	411	-	-	80	11,23,200.00
	Grand Total	3,171	14,405 528	1154.10		2643	3,88,85,103.00

Annexure-IV
Village-wise Population Affected

S.No.	Name of village.	No. of families displaced				Total population.
		S.C.	S.T.	General	Total	
1.	Semla	2	-	12	14	75
2.	Kechela	11	118	42	171	755
3.	Thalaput	14	23	37	74	375
4.	Koranga	69	218	35	322	1288
5.	Puki	33	72	8	113	565
6.	Paidaput	30	6	29	65	325
7.	Malipondi	-	-	60	60	345
8.	Parojapondi	19	196	12	227	990
9.	Putkoranga	24	34	-	58	322
10.	Sirisi	11	1	63	75	375
11.	Bhoiguda	-	42	-	42	210
12.	Kalchuru	3	51	10	64	256
13.	Thataguda	-	2	22	24	96
14.	Kulaba (2nd phase displacement)	11	27	114	152	608
15.	Chandalguda	-	8	15	23	115
16.	Gatanput	-	7	-	7	35
17.	Landiguda	9	38	31	78	390
18.	Dasarikhariguda	7	58	7	72	360

(Contd.)

Annexure-IV (Contd.)

19.	Tiadiput (2nd phase displa- cement	3	1	91	95	290
20.	Tiadiput (4th phase displa- cement)	6	-	20	26	134
21.	Gopalput	8	103	25	141	424
22.	Machaput	10	46	-	56	280
23.	Balda (Routput) (2nd phase dis- placement)	-	57	56	113	384
24.	Balda (Doraput) (3rd phase dis- placement)	30	-	33	63	320
25.	Sindhiput (Nandapur P.S.)	1	-	26	27	58
26.	Malipeta	6	3	99	108	800
27.	Dhadi	31	81	35	147	735
28.	Ghataguda	-	-	1	1	5
29.	Mohanpada	-	-	2	2	9
30.	Teraguda	-	18	-	18	82
31.	Mundaguda	-	5	-	5	25
32.	Khamari	1	4	34	39	195
33.	Dangaguda (Masiput)					
34.	Masiput	-	9	6	15	95

(Contd.)

Annexure-IV (Contd.)

35.	Dumuriguda	2	-	37	39	400
36.	Jamuguda 3rd phase	-	-	12	12	60
	(Kulaba) displacement	-	-	3	3	15
37.	Andalaguda (Kulaba)					
38.	Haldiput	3	-	26	29	150
39.	Sargunaput	1	12	3	16	85
40.	Lanjiguda (Khalapadi)	-	34		34	136
41.	Kuturiput (3rd phase) displacement	5	3	16	24	95
42.	Pakhanput (4th phase dis-placement)	-	7	-	7	29
43.	Tentulipada (3rd phase displacement)	11	-	-	11	32
44.	Tentulipada (4th phase displacement)	8	-	-	8	30
45.	Jaraput (Rajubedei)	-	-	18	18	95
46.	Chikapar	43	79	43	165	460
47.	Chagam	-	-	1	1	4

(Contd.)

Annexure-IV (Contd.)

48.	Kaki	-	-	2	2	9
49.	Sindhiput (Machkund PS)	7	-	13	20	105
50.	Kadrimaliput	2	-	17	19	95
51.	Biriguda	-	-	28	28	102
52.	Khapraput	-	-	3	3	22
53.	Khilaput (Koraput)	-	-	21	21	84
54.	Charaguda	2	8	-	10	55
55.	Godiput	-	-	8	8	40
56.	Khilaput (Lamptaput)	-	-	1	1	4
57.	Mahadeiput	-	-	11	11	50
58.	Badaliguda hamlet	15	45	20	80	411
59.	Dhemsaguda hamlet					
60.	Khapraput & Khilaput (shifted in 1991)	4	-	100	104	516
	Total	442	1421	1308	3171	14405

Annexure-V

Abstract of Revised Estimate of Upper Kolab Project-Sub-head "Reservoir" from which L.A. and Rehabilitation Expenditure is met-1987.

S.No.	*Item of work*	*Amount (Rs. in lakh)*
1.	Preliminary (Survey & Investigation).	24.00
2.	Compensation for private land.	689.11
3.	Compensation for forest land and waste land - 6556 acres @ Rs.360 per acre.	23.60
4.	Land for colonies at the dam site and other project sites.	4.50
5.	Realignment of roads.	515.89
	(a) N.H.43	73.55
	(b) M.D.R.52 - 182/0 to 186/0 Dabuguda 169/6 to 174/4 Tusuba	34.56
	(c) M.D.R.55 (Pithaguda bridge & approaches)	104.63
6.	Reconstruction of Kanyashram building at Deoghati.	11.17
7.	Compensation to P.H.D.	152.40
	(i) Jawahar Water Works, Sunabeda.	20.67
	(ii) Koraput Water Works	
8.	Construction of roads and bridges at Central Cattle Breeding Farm, Semiliguda.	20.00
9.	Additional cost compensation for marooned villages 2451.21 acres @ Rs.2500/- per acre.	61.28

(Contd.)

Annexure-V (Contd.)

10.	Payment to Orissa State Electricity Board, Orissa Small Industries Corporation & P and T Department and other Departments.	46.91
11.	Payment to Forest Department for compensatory afforestation.	10.00
12.	Provision for sedementation study.	5.00
13.	Rehabilitation of displaced persons.	
	(a) Survey & investigation of colonies.	7.50
	(b) Providing common facilities.	40.00
	(c) Transporation of displaced persons.	30.00
	(d) Development and reclamation of land.	100.00
	(e) Land acquisition for roads, boundary pillars etc.	3.00
	(f) Buildings for Resettlement & Rehabilitation Officer's establishment.	10.00
	(g) Two nos. of High School Buildings.	12.50
	(h) Rehabilitation cash grant @ Rs.14,040/- per family for 3000 families.	421.20
	(i) Rehabilitation cash grant for 300 families of marooned villages @ Rs.14,040/- per family.	42.12
14.	Provision for soil conservation measures in the catchment area.	500.00
		2963.59
		or say Rs.2964.00 lakh

Annexure-VI

(Format to Obtain Informations about Oustees Resettled at Different Places).

I. Name of the oustee.
   - (*a*) Name of the father/husband
   - (*b*) Age
   - (*c*) Caste
   - (*d*) Educational qualification
   - (*e*) No. of members in family (relationship with the settler & their age).

II. Prior to displacement
   - (a) (i) Name of village from which displaced.
     - (ii) Police Station
     - (iii) Tehsil
     - (iv) Block
     - (v) Name of nearby market place and its distance.
   - (b) No. of families (caste-wise)
   - (c) Population (caste-wise)
   - (d) No. of families affected by the Project.
     - (i) Fully affected.
     - (ii) Partly affected.
   - (e) Total area of the village and the area out of it, affected:-
     - (i) Agricultural land.
     - (ii) Forest land.
     - (iii) Others.

III. Civic amenities existing in the village before submergence.
   - (a) Roads
   - (b) School (Standard)

(Contd.)

Annexure-VI (Contd.)

(c) Wells
(d) Tanks
(e) Dispensary/Hospital
(f) Club House
(g) Post Office
(h) Electricity & others.

IV. Previous occupation

V. Approximate annual income.

VI. (a) Approximate cost of property lost due to submergence.
(i) Extent of Homestead land (ii) No. of rooms (iii) Classification of House (Whether pucca, or thatched etc) (iv) Cost of land & property.
(b) Agricultural land.
(a) Extent (b) Classification (c) Irrigation facility available or not (d) Cost of land.
(c) Trees.
(i) Type of tree & their number
(ii) Cost
(d) Others.

VII. Benefits received after displacement.
a) Compensation money if any & its amount.
b) Agricultural land (its extent, classification, approximate cost and irrigable or not)
c) Employment (category, name of employer, monthly salary).
d) Vocational training if any.
e) Transport charges.
f) House building materials etc.

VIII. Place of resettlement after displacement. (Name of village, whether Revenue village or Settlement colony, Police Station, Tehsil, Block, Market Place nearby and its distance).

(Contd.)

Annexure-VI (Contd.)

IX. If resettled in settlement colony.
  (a) Total No. of families resettled there (caste-wise)
  (b) Population (Caste-wise)
  (c) Civic amenities available - (Roads, wells, tanks, schools, Dispensary, Club House, Panchayat Ghar, Temple, others).

X. Present occupation and approximate annual income.

XI. Whether able to maintain livelihood or not.

XII. Views of the resettler on the civic amenities, society, etc. in the new environment.

Annexure-VII
(Format to Collect Informations Relating to Submerged Villages).

1. Name of village
2. No. of families residing
3. Population
4. Date of notification U/S 4(1) of L.A. Act, 1894.
5. Date of payment of compensation money.
6. Public institutions affected - (Schools, Temple, Club House, Panchayat Ghar, Dispensaries & others).
7. Relocation of Public Institutions and the place where relocated.

Annexure-VIII
(Format to Collect Informations Relating to Settlers in Colony No.7).

1. Name of the settler
2. Colony number
3. Annual income:

(a) In the old village	*(b) In the Settlement Colony*
(i) From Agriculture	(i) From Agriculture.
(ii) From service	(ii) From service.
(iii) Other sources	(iii) Other sources

4. Agricultural implements in possession:

Items	*In the old village*	*In the settlement colony*
(a) Plough		
(b) Sickle		
(c) Bullock cart		
(d) Axex		
(e) Spade		
(f) Others		

5. Livestock position:

Items	*In the old village*	*In the Settlement colony*
(a) Cow		
(b) Bullock		
(c) Buffaloe		
(d) Goat		
(e) Sheep		
(f) Poultry		

6. Manner of utilisation of compensation money/Rehabilitation grant received.

Bibliography

Bibliography

1. Anthony Oliver Smith, Department of Anthropology, University of Florida, U.S.A.- "Involuntary Resettlement and Political empowerment" - a paper originally presented at the International Research and Advisory Panel, Refugee, Studies Programme, university of Oxford, January, 1991.
2. Bandyopadhyay J., Jayal N.D., Schoettli U & Singh Chhatrapati - "India's Environment - Crises and responses", Natraj Publishers, Rajpur Road, Dehra Dun.
3. Behura, N.K. Dr., Professor & Head, Department of Anthropology, Utkal University - "Socio-economic problems and Social change among the relocatees of Rengali Dam - a case study;" article published in "Man and Life", Vol.15 - Nos. 1&2, January-June, 1989.
4. Bell R.C.S., I.C.S. - "Orissa District Gazetteers - Koraput", 1945.
5. Butcher David A.P. - An Operation Manual for Resettlement Rome, Food & Agriculture Organisation of the United Nations, 1971.
6. Cernea Michael M - "Involuntary Resettlement in Development Projects" - Policy guidelines in World Bank financed Projects, World Bank Technical Paper No. 80.
7. Chambers Robert - "The Volta Resettlement Experience" - Palm Mall Press, London, 1970.
8. "Circulars and Orders on Rehabilitation and Resettlement of Rengali Multipurpose Project" - Published by Irrigation & Power Department, Government of Orissa.
9. Circulars and orders issued by different State Governments:-
   (a) Andhra Pradesh
      (i) Memo No.560/Ir. VIII-86 dt. 25.6.1986.
      (ii) G.O.Ms No.59 Dt. 19.3.1980.
      (iii) G.O.Ms No.460 dt. 27.12.1983
      (iv) G.O.Ms No.147 dt. 2.4.1981
      (v) G.O.Ms No.98 Dt. 15.4.1986.

(b) Bihar - "Rehabilitation and Resettlement Plan" of Subarnarekha Multipurpose Project.

(c) Gujarat - "Sardar Sarovar Project - Land Acquisition and Rehabilitation of oustees."

(d) Karnataka - "Action Plan to rehabilitate project affected families in the Upper Krishna Project"; prepared by MYRADA, 2, Service Road, Domlur Layout, Bangalore.

(e) Madhya Pradesh - "Rehabilitation Policy of Madhya Pradesh Government in respect of families affected by submergence due to Narmada Sagar Complex Projects."

(f) Tamilnadu-

(i) G.O.Ms No.202 Dt. 25.7.1986.

(ii) G.O.Ms No.354 Dt. 21.3.1986.

(iii) G.O.Ms No.188 Dt. 28.12.1976.

10. Circulars and orders issued by Government of India.

(a) Letter No.27(9)80-PI. Dt.19.5.1980 of the Ministry of Energy & Irrigation, Department of Irrigation.

(b) Office Memorandum No.15/13/84-BPE(C) Dt.3.2.1986, Ministry of Industry, Department of Public Enterprises, Bureau of Public Enterprises.

(c) Report of the Working Group on development of scheduled tribes during 7th Five-Year Plan, Published by the Ministry of Home Affairs.

11. "Committee on Estimates" - 1985-86, 34th and 37th Report on rehabilitation of displaced persons of Rengali Multipurpose River Dam Project - Published by the Secretariat of the Orissa Legislative Assembly.

12. Constitution of India.

13. Dalua, A.K - "Irrigation in Orissa" - A compilation under publication by Irrigation Department, Government of Orissa.

14. Das, R.C. Dr. Chairman, State Prevention and Control of Pollution Board, Orissa - "State of Environment in Orissa-1-Water Resources."

15. Doria, R.S., Singh Avadesh K, Singh Surendra - "Man, Development and Environment." Ashish Publishing House, 8/81, Punjabi Bagh, New Delhi.

16. "Draft Project Outline" prepared by the Orissa Remote Sensing Application Centre, Department of Science, Technology & Environment, Government of Orissa.

17. "Draft report on rehabilitation and resettlement of families displaced in the irrigation projects - Ramial, Kuanria and Talsara".

Prepared by M/s Project and Corporate Consultants Pvt Ltd., Bhubaneswar.

18. "Economic Survey, 1988-89", published by Planning & Coordination Department, Government of Orissa.

19. "Eco Development of Eastern Ghats" - prepared by the Institute for Coastal and Offshore Research, Vishakapatnam for Department of Energy, Environment, Science & Technology, Government of Andhra Pradesh.

20. Engineer Rusi - "The Sardar Sarovar Controversy - Are the critics right ?" - Special report published in the Business India, October 30 to November 12,1989.

21. Environment (Protection) Act, 1986.

22. Forest (Conservation) Act, 1980.

23. Goldsmith E, and Hildyard N - "The Social and Environmental Effects of Large Dams (2 Vols.) European Ecological Action Group and Wade Bridge Ecological Centre, Camelford, Cornwall, 1984.

24. Gosling, Peter L.A. - "Population Resettlement in the Mekong river basin" - Papers of the third Carolina Geographical Symposium (editor), University of North Corolina at Chapel Hill, Department of Geography, Chapel Hill, NC 27514.

25. "Guidelines for Aid Agencies in involuntary displacement and Resettlement in Development Projects - Published by Organisation for economic co-operation and Development, Paris.1991.

26. "History of Upper Kolab Project" Vol I, Published by Government of Orissa, Irrigation & Power Department.

27. Guggenheim Scott - Environment Department, World Bank - "Development and the Dynamics of Displacement".

28. Joshi, B.L. Dr., "Displacement and Rehabilitation", Parimal Prakashan, Aurangabad, Maharashtra.

29. Land Acquisition Act, 1894.

30. Maharashtra Act No. XLI of 1976 - "The Maharashtra Resettlement of Project Displaced Persons Act, 1976".

31. Mishra, R.K. I.A.S. Chief Secretary, Orissa - "Environmental Planning and Forest Management."

32. Murty, S.S. I.A.S, Deputy Commissioner and incharge of Hirakud Land Organisation, Government of Orissa - "Resettlement and Reclamation at Hirakud."

33. "National Water Policy", published by the Government of India, Ministry of Water Resources, 1987.

34. Nayak, P.K.,Reader in Anthropology, Utkal University, "Displaced denisens of Rengali Dam Project - A study of their socio-economic

conditions, Resettlement problems, and ameliorative action measures."

35. "Orissa Government Land Settlement Act, 1962."

36. "Our Environment - our future," published by the Ministry of Environment & Forest, Government of India.

37. Partridge W., Resettlement Specialist of World Bank - "Suggested outline and recommendations for an action plan for resettlement and rehabilitation in the Upper Indravati Project" in Orissa.

38. Partridge, W.L.; Brown, A.B. and Nugent, J.B. - "The Papaloapan Dam and Resettlement Project. Human Ecology and Health Impact". In involuntary migration and resettlement. A Hansen and A.Oliver Smith, editors, pp.-245-263. Boulder, Colorado West View Press.

39. Proceedings of the meetings of Rehabilitation Adivosry Committee of Upper Kolab Project.

40. "Problems in Rehabilitation of displaced tribals in Upper Indravati and Rengali Multipurpose Project - a Study conducted by Operation Research Group and sponsored by Planning Commission, Government of India.

41. Proceedings of Western Regional Seminar, Southern Regional Seminar, Northern Regional Seminar and Eastern Regional Seminar on "Water Resources Development - Environmental issues and Solutions" held at Ahmedabad, Trivandrum, Bhopal and Bhubaneswar respectively in 1987 and published by Water and Power Consultancy Services (India) Ltd., New Delhi.

42. Raj, P.A. - Vice Chairman and Managing Director, Sardar Sarovar Narmada Nigam Ltd., Gandhinagar (Gujarat) - "FACTS - Sardar Sarovar Project."

43. Ramiah, P. - "Tribal Economy of India" - Published by Light and Life Publishers, New Delhi.

44. Ranade, V.M, Chief Engineer, Irrigation Department, Govt. of Maharashtra - "Reservoirs and Environment."

45. Recommendations of the National Seminar on Water Resources Development - Environmental issues and solutions held on 29th February, 1988, at Vigyan Bhavan, New Delhi.

46. Records of Irrigation Department.

47. Reddy, I.U.B. Faculty Member, Tata Institute of Social Sciences, Bombay - "Impact of large irrigation projects on the tribals", published in "Social Change", March 1988, Vol 18. No.1.

48. Reddy, I.U.B.Faculty Member, Tata Institute of Social Sciences, Bombay, "Costs of River Valley Projects in India", published by the Indian Institute of Public Administration, New Delhi.

49. Report on the Study of Rehabilitation of tribals displaced by Upper Kolab Project in Koraput district and NALCO Project in Dhenkanal district - Govt. of India, Office of the Director for Scheduled Castes & Scheduled Tribes, Bhubaneswar.

50. Sawant, Prakash R., Professor., Department of Geography, Kisan Veer Mahavidyalaya, Wai, Maharashtra - "River Dam construction and resettlement of affected villages" - a case study of two villages in the Upper Krishna Valley, Maharashtra.

51. Schudder, Thayer, 1973 - "The human ecology of big project: river basin development and resettlement." Annual review of Anthropology, Vol. 2, 45-61.

52. Schudder Thayer and Elizabeth F. Colson,1982 - "From Welfare to development: a conceptual frame work for the analysis of dislocated people" - A.Hansen and A.Oliver Smith on Involuntary migration and Resettlement - The Problems and responses of dislocated prople.

53. Sharma, B.D. Dr., Commissioner for S.C & S.T., June 1986 - "Rehabilitation of tribals displaced by irrigation projects - An ecologically viable and socially compatible alternative approach."

54. "Statistical Abstract of Orissa,1981", Published by Bureau of Statistics & Economics,Orissa.

55. Sudarsen V. and M.A.Kalam - "The Uprooted-Development, Displacement and Resettlement" - Gian Publishing House, New Delhi.

56. Sundaram, K.V.- "Urban and Regional Planning in India," Published by Vikas Publishing House Pvt Ltd, New Delhi.

57. Verma, R.P. D.I.C.,M.Sc (London) D.S.C (M.I.T), S.E., Development Circle, Hirakud Dam Project, Orissa - "Hirakud Reservoir and its Problems."

58. Verma, S.C., Chairman, N.V.D.A - "Human Resettlement in Lower Narmada Basin", Published by Narmada Valley Development Authority, Government of Madhya Pradesh, Bhopal.

INDEX